RAND

Science and Technology Policy Institute

The National Bioethics Advisory Commission

CONTRIBUTING TO PUBLIC POLICY

ELISA EISEMAN

Prepared for the
National Bioethics Advisory Commission

The research described in this report was conducted by RAND's Science and Technology Policy Institute for the National Bioethics Advisory Commission.

Library of Congress Cataloging-in-Publication Data

Eiseman, Elisa.
    The National Bioethics Advisory Commission : contributing to public policy /
Elisa Eiseman.
        p. cm.
        "MR-1546."
        Includes bibliographical references.
        ISBN 0-8330-3364-6 (pbk. : alk. paper)
        1. Bioethics—Government policy—United States. 2. Medical ethics—
Government policy—United States.
        [DNLM: 1. United States. National Bioethics Advisory Commission. 2.
Biomedical Research. 3. Bioethical Issues. 4. Biomedical Research. 5. Cloning,
Organism. 6. Human Experimentation. 7. Public Policy. W 20.5 E36n 2003] I.
Rand Corporation. II.Title.

QH332.E36 2003
174'.957'0973—dc21

2003008549

RAND is a nonprofit institution that helps improve policy and decisionmaking through research and analysis. RAND® is a registered trademark. RAND's publications do not necessarily reflect the opinions or policies of its research sponsors.

About the cover: The picture portrays nature, science, and the impact of man on both. "DNA Nucleotides between Dogwood & the Judas Tree," stoneware wall plaque, by Jane W. Larson, Bethesda, MD. Photograph by Philippe C. Bishop.

Published 2003 by RAND
1700 Main Street, P.O. Box 2138, Santa Monica, CA 90407-2138
1200 South Hayes Street, Arlington, VA 22202-5050
201 North Craig Street, Suite 202, Pittsburgh, PA 15213-1516
RAND URL: http://www.rand.org/
To order RAND documents or to obtain additional information, contact Distribution Services: Telephone: (310) 451-7002;
Fax: (310) 451-6915; Email: order@rand.org

The National Bioethics Advisory Commission (NBAC) was established by Executive Order 12975 in October 1995 to advise the National Science and Technology Council and other appropriate government entities regarding "bioethical issues arising from research on human biology and behavior." NBAC was established in response to proposals by the National Institutes of Health (NIH), the Department of Energy, and other research-oriented agencies; recommendations of the Advisory Committee on Human Radiation Experiments (ACHRE); and a recognized need by the White House Office of Science and Technology Policy (OSTP) for a national commission to address a broad set of ethical issues, including genetic privacy and the protection of human research subjects. NBAC met for the first time on October 4, 1996. On October 3, 2001, the commission's charter expired, and NBAC's tenure ended. Over that five-year period, NBAC met 48 times and submitted six major reports to the White House. NBAC's six reports covered topics ranging from cloning human beings and stem cell research to research with human biological materials and the protection of human participants in research.

As part of an Intergovernmental Personnel Act (IPA) agreement between RAND and NBAC, the author of this report tracked the response to the six NBAC reports and the recommendations they contained from March 2000 until October 2001 when NBAC's charter expired. This study was conducted in an effort to assess NBAC's contribution to the policymaking process as it relates to various bioethical and scientific issues. This report describes these documents and provides a detailed account of the response to each of NBAC's six reports by the White House, Congress, federal and state governments, professional societies, philanthropic foundations, organizations representing private industry, patient advocacy groups, other countries, and international organizations. Because these entities are responsible for decisionmaking and policy formulation in this country and abroad, their response to NBAC's reports and recommendations can be used as an indication of the extent to which the commission's work is reflected in public policy and public discourse.

Even though NBAC's tenure has concluded, debate on many of the issues addressed by NBAC has continued and will continue for some time to come. Therefore, this report represents an early assessment of the response to NBAC's work. Given the complex nature of the issues addressed by NBAC and the pace of the policymaking process, the response to NBAC's reports and recommendations should be assessed over time to determine both their immediate and long-term impact.

This report should be of value to anyone interested in the role of commissions such as NBAC in the policymaking process. This report should also be of value to anyone interested in bioethical issues, especially the ethical use of new scientific discoveries and technological innovations and the protection of human participants in research. Policymakers, the research community, and the public should be interested in what NBAC recommended and the response to its work.

The tracking and analysis of documents referring to or based on NBAC's work and the draft version of this report were done primarily while the author was working with NBAC through the IPA agreement. When NBAC's charter expired, the author returned to the RAND Science and Technology Policy Institute and completed work on this document. Completion of this project was funded jointly by NBAC and the Science and Technology Policy Institute (S&TPI).

ABOUT THE WHITE HOUSE OFFICE OF SCIENCE AND TECHNOLOGY POLICY

OSTP was created in 1976 to provide the president with timely policy advice and to coordinate the federal investment in science and technology. OSTP's Technology Division helps to develop and implement federal policies for harnessing technology to serve national goals, such as global economic competitiveness, environmental quality, and national security. The Technology Division's priorities include sustaining U.S. technological leadership through partnerships to promote the development of innovative technologies, promoting research and development and policy initiatives for advanced computing and communications technologies, advancing technologies for education and training, and promoting the U.S. space and aeronautics program.

ABOUT THE SCIENCE AND TECHNOLOGY POLICY INSTITUTE

Originally created by Congress in 1991 as the Critical Technologies Institute and renamed in 1998, the Science and Technology Policy Institute is a federally funded research and development center sponsored by the National Science Foundation and managed by RAND. S&TPI's mission is to help improve public policy by conducting objective, independent research and analysis on policy issues that involve science and technology. To this end, S&TPI

- supports the Office of Science and Technology Policy and other Executive Branch agencies, offices, and councils

- helps science and technology decisionmakers understand the likely consequences of their decisions and choose among alternative policies

- helps improve understanding in both the public and private sectors of the ways in which science and technology can better serve national objectives.

In carrying out its mission, S&TPI consults broadly with representatives from private industry, institutions of higher education, and other nonprofit institutions.

Inquiries regarding the Science and Technology Policy Institute may be directed to:

Helga Rippen, Director
RAND Science and Technology Policy Institute
1200 South Hayes Street, Arlington, VA 22202-5050
rippen@rand.org

CONTENTS

TABLES

BACKGROUND ON THE NATIONAL BIOETHICS ADVISORY COMMISSION

The National Bioethics Advisory Commission (NBAC) was established by Executive Order 12975 in October 1995 to advise the National Science and Technology Council and other appropriate government entities regarding "bioethical issues arising from research on human biology and behavior." NBAC was established in response to an unmet need for a national commission to address a broad set of ethical issues. NBAC was the fifth federal bioethics commission created to contribute to the development of public policy and to promote a national discussion on complex bioethical issues. NBAC met for the first time on October 4, 1996. On October 3, 2001, the commission's charter expired, and NBAC's tenure ended. During its tenure, NBAC met 48 times and submitted six major reports to the White House.

The NBAC reports offered 120 recommendations designed to improve the protection of human participants in research while also supporting the continued advancement of science and the promotion of ethically sound research. Three of the reports were requested by the president—*Cloning Human Beings, Ethical Issues in Human Stem Cell Research*, and *Ethical and Policy Issues in Research Involving Human Participants*—and three were initiated by the NBAC commissioners—*Research Involving Persons with Mental Disorders That May Affect Decisionmaking Capacity, Research Involving Human Biological Materials: Ethical Issues and Policy Guidance*, and

Ethical and Policy Issues in International Research: Clinical Trials in Developing Countries.

THE RESPONSE TO NBAC'S REPORTS AND RECOMMENDATIONS

This report provides a detailed account of the responses to each of NBAC's six reports by the president, U.S. Congress, federal and state government, professional societies, philanthropic foundations, organizations representing private industry, patient advocacy groups, other countries, and international organizations. These entities are responsible for decisionmaking and policy formulation in regard to scientific issues in this country and abroad. Therefore, their response to NBAC's reports and recommendations can be used as an indication of the extent to which NBAC's work is reflected in public policy and public discourse.

The response was tracked by collecting documents that referred to or were based on any of NBAC's reports and recommendations, including but not limited to federal and state legislation; congressional testimony; presidential administration and federal agency guidelines, statements, policies, and procedures; statements from various professional and academic societies, organizations, and foundations; and laws, regulations, and guidelines from other countries and international organizations. Discussions on NBAC's reports and recommendations in the academic literature and in the media were also tracked.

Whether or not policy recommendations are implemented depends in part on how those recommendations are formulated and to whom they are targeted. NBAC took this into account when it was formulating its recommendations by carefully considering the types of policy changes it was recommending and to whom they were directed. For example, NBAC made several recommendations directed to Congress, state governments, and federal agencies that called for a variety of actions, including legislation, regulations, guidance documents, and support for both professional and public education. NBAC also made several recommendations to the scientific and medical communities, a number of which were directed specifically to professional societies, funding institutions, and the private sector, and appealed to the private sector for voluntary compliance with many of its recommendations. Furthermore, in some of its reports, NBAC recommended international cooperation.

NBAC's reports and recommendations prompted responses from the president of the United States and the presidential administration, Congress, state governments, federal agencies, professional societies, philanthropic foundations, organizations representing private industry, patient advocacy groups, other countries, and international organizations. The responses varied from a passing mention of NBAC to the introduction of state and federal legislation based on NBAC's recommendations and adoption of NBAC's recommendations into guidance and policy statements. However, not all of the responses to NBAC's reports and recommendations were favorable. Some of its recommendations were criticized as being cumbersome, impractical, and in danger of impeding valuable research, while some were criticized for not being restrictive enough.

Government Response

Several types of policy were implemented in response to NBAC's recommendations. Although no federal or state legislation was signed into law, federal and state bills based on NBAC's recommendations were introduced, and Congress was informed about NBAC's work through congressional testimony by NBAC commissioners and others. Thirteen bills were introduced in Congress that mentioned NBAC—four were introduced in response to the human cloning debate, eight dealt with the privacy of genetic and medical information, and one addressed the system for oversight for the protection of human research participants. In addition, bills on human cloning, decisionmaking capacity, and genetic privacy that mentioned NBAC or were based on NBAC's recommendations were introduced in the state legislatures of four states. In addition, NBAC commissioners and staff testified before Congress 18 times after being invited by subcommittees of the House and Senate to discuss ongoing work and completed projects.

Federal agencies have issued guidance and policy statements in response to NBAC's reports and recommendations. The National Institutes of Health (NIH), the government agency responsible for the majority of federally funded research involving human participants, has issued guidance based on recommendations made in three of the NBAC reports—*Research Involving Persons with Mental Disorders That May Affect Decisionmaking Capacity*, *Research Involving Human Biological Materials*, and *Ethical Issues in Human Stem Cell Research*. The Food and Drug Administration (FDA) has adopted NBAC's terminology and recommendations on informed consent found in the *Research Involving Human Biological Materials* report, and the National Institute for Occupational Safety and Health (NIOSH) at the Centers for Disease Control and Prevention (CDC) has instructed all investigators who perform research involving persons with developmental disabilities to review NBAC's recommendations on informed consent in the *Research Involving Persons with Mental Disorders That May Affect Decisionmaking Capacity* report.

Response from Professional Societies, Organizations, and Foundations

A number of professional and academic societies, organizations, and foundations have taken note of NBAC's reports and have released policy statements, issued guidance documents, and developed educational materials based on NBAC's recommendations for their members to consider. Several professional and scientific societies, as well as organizations representing the biotechnology and pharmaceutical industries, have called for a five-year voluntary moratorium on human cloning, as recommended in the *Cloning Human Beings* report. The Alzheimer's Association funded research on the potential impact of NBAC's recommendations in the *Research Involving Persons with Mental Disorders That May Affect Decisionmaking Capacity* report. Intermountain Health Care issued policies and procedures for institutional review boards (IRBs), and the Online Ethics Center for Engineering and Science developed an online teaching module, both of which were based on NBAC's recommendations in *Research Involving Human Biological Materials*. The Endocrine Society in its Code of Ethics and the American Medical Association (AMA) in official policy supported the NBAC's *Ethical Issues in Human Stem Cell Research* report and the recommendations in it.

International Response

Other countries and several international organizations have studied NBAC's reports and have supported some of the commission's recommendations. The United Kingdom's Human Fertilisation and Embryology Authority (HFEA) and the Human Genetics Advisory Commission (HGAC) supported NBAC's recommendations in *Cloning Human Beings*, and the Nuffield Council on Bioethics endorsed some of the recommendations in *Ethical Issues in Human Stem Cell Research*. NBAC's reports and recommendations have been cited in various publications in Australia, Canada, Japan, and the United Kingdom and in reports by several international organizations, including the Organisation for Economic Co-operation and Development (OECD) and the United Nations Educational, Scientific, and Cultural Organization (UNESCO). In addition, the NBAC reports have been widely circulated internationally, and have been translated into Japanese and published in a German science and ethics journal.

SHORT-TERM AND LONG-TERM ASSESSMENT OF NBAC'S WORK

Even though NBAC's tenure has concluded, debate on many of the issues addressed by NBAC has continued, and will continue for some time to come. Therefore, this report represents an early assessment of the response to NBAC's work. However, given the complex nature of the issues addressed by NBAC and the usual pace of the policy-making process, the response to NBAC's reports and recommendations should also be assessed over time to determine both their immediate and their long-term impact.

Succeeding administrations and bioethics commissions, and in particular the President's Bioethics Council, should be very interested in what NBAC recommended and the response to its work. During its deliberations, NBAC examined the work of several previous commissions, including the National Commission for the Protection of Human Subjects of Biomedical and Behavioral Research, the President's Commission for the Study of Ethical Problems in Medicine and Biomedical and Behavioral Research, and the Advisory Committee on Human Radiation Experiments.

NBAC has increased the awareness of the U.S. and foreign governments, international groups, the research community, and the public
about complex bioethical issues, providing a forum for their public
debate, and making recommendations that have been incorporated
into the system of oversight for the protection of human research
participants. Several of the issues that were addressed by NBAC in its
reports, including human cloning, human stem cell research, and the
protection of human research participants, are currently under discussion or are being examined by Congress, the President's Council
on Bioethics, some federal agencies, and various other groups for
legislation, recommendations, or other actions. Therefore, NBAC's
reports and recommendations will continue to be relevant to ongoing policy debates on bioethics issues.

ACKNOWLEDGMENTS

The author would like to thank a number of people who contributed to this report:

Eric Meslin, Ph.D., former Executive Director of NBAC, helped to establish an ongoing relationship between RAND and NBAC and gave the author the opportunity to work closely with NBAC for over four and a half years. Dr. Meslin conceived of and initiated the project to track the response to NBAC's reports and recommendations and reviewed the final draft version of this report.

Debra McCurry, M.S., former Information Specialist at NBAC who was an integral part of this project, performed literature searches and analyses of the academic literature and the media for Chapter Two of this report, "NBAC in the Public Eye," and provided valuable comments on early drafts of this report.

Marjorie Speers, Ph.D., former Acting Executive Director of NBAC, supported the project and provided valuable comments on early drafts.

Glen Drew, M.S., J.D., former Senior Research Policy Analyst at NBAC, provided information on the international response to NBAC's reports and recommendations and provided valuable comments on early drafts.

Ellen Gadbois, Ph.D., former Senior Policy Analyst at NBAC, provided information on the government response to NBAC's reports and recommendations and provided valuable comments on early drafts.

Kerry Jo Lee, former Program Analyst at NBAC, provided information on the society, organization, and foundation response to NBAC's reports and recommendations.

Dan Pastor, former summer intern at NBAC, prepared an annotated bibliography of media and academic literature critiques of NBAC's first two reports—*Cloning Human Beings* and *Research Involving Persons with Mental Disorders That May Affect Decisionmaking Capacity.*

Richard A. Rettig, of RAND, reviewed the final draft version of this document.

Gail Kouril, of RAND, performed Lexis-Nexis searches for state legislation that mentions NBAC.

Lisa Sheldone, of RAND, provided administrative expertise that was essential in bringing this project to fruition.

Nancy DelFavero, of RAND, edited the final report.

Other former NBAC research staff contributed insight, information, and support for this project:

Doug Berger, M. Litt.   Kathi Hanna, M.S., Ph.D.
Stu Kim, J.D.   Anne Drapkin Lyerly, M.D.
Ayodeji Marquis   Alice Page, J.D., M.P.H.
Robert Tanner, J.D.

The following former NBAC administrative staff members provided assistance and support on this project:

Nicole Baker   Jody Crank
Evadne Hammett   Margaret Quinlan
Sherrie Senior

The author also thanks the following former NBAC commissioners, whose reports and recommendations that are the subject of this report contributed to the public policymaking process:

Harold T. Shapiro, Ph.D., Chair   Patricia Backlar
Arturo Brito, M.D.   Alexander Morgan Capron, LL.B.
Eric J. Cassell, M.D., M.A.C.P.   R. Alta Charo, J.D.
James F. Childress, Ph.D.   David R. Cox, M.D., Ph.D.

Rhetaugh Graves Dumas, Ph.D., R.N.

Carol W. Greider, Ph.D.

Bette O. Kramer

Lawrence H. Miike, M.D., J.D.

William C. Oldaker, LL.B.

Laurie M. Flynn

Steven H. Holtzman

Bernard Lo, M.D.

Thomas H. Murray, Ph.D.

Diane Scott-Jones, Ph.D.

And finally, the author wishes to thank everyone else who provided support and valuable information on the response to NBAC's work and its contribution to public policy.

ACRONYMS

AAAS	American Association for the Advancement of Science
AAMC	Association of American Medical Colleges
AAN	American Academy of Neurology
ACD	Advisory Committee to the Director, National Institutes of Health
ACHRE	Advisory Committee on Human Radiation Experiments
AHEC	Australian Health Ethics Committee
AHRP	Alliance for Human Research Protection
AIDS	Acquired immunodeficiency syndrome
AMA	American Medical Association
ANA	American Neurological Association
ASIP	American Society for Investigative Pathology
BIO	Biotechnology Industry Organization
BMA	British Medical Association
CBAC	Canadian Biotechnology Advisory Committee
CBER	Center for Biologics Evaluation and Research
CDC	Centers for Disease Control and Prevention
CEJA	(American Medical Association) Council on Ethical and Judicial Affairs
CIHR	Canadian Institutes of Health Research

CIOMS	Council for International Organizations of Medical Sciences
CLIA	Clinical Laboratory Improvement Amendments
CMO	Chief Medical Officer (United Kingdom)
COPR	(Director's) Council of Public Representatives
CST	Council for Science and Technology
DAIDS	Division of AIDS
DHHS	Department of Health and Human Services
DMID	Division of Microbiology and Infectious Diseases
DNA	Deoxyribonucleic acid
EGE	European Group on Ethics in Science and New Technologies
FASEB	Federation of American Societies for Experimental Biology
FDA	Food and Drug Administration
FIC	Fogarty International Center (for Advanced Study in the Health Sciences)
HFEA	Human Fertilisation and Embryology Authority
HGAC	Human Genetics Advisory Commission
HHS	Health and Human Services
HIPAA	Health Insurance Portability and Accountability Act
HIV	Human immunodeficiency virus
HRSA	Health Resources and Services Administration
HSRAC	Human Subjects Research Advisory Committee
HSRCW	Human Subject Research Council Workgroup
HUGO	Human Genome Organisation
IBC	International Bioethics Committee (of the United Nations Educational, Scientific and Cultural Organization)
ICS	Institute for Civil Society
IHC	Intermountain Health Care

IHS	Indian Health Service
IPA	Intergovernmental Personnel Act
IRB	Institutional review board
JSC	Joint Steering Committee (for Public Policy)
JPMA	Japan Pharmaceutical Manufacturers Association
LAR	Legally authorized representative
NACDA	National Advisory Council on Drug Abuse
NAMHC	National Advisory Mental Health Council (to NIMH)
NANDS	National Advisory Neurological Disorders and Stroke (Council)
NBAC	National Bioethics Advisory Commission
NCHSTP	National Center for HIV, STD and TB Prevention
NHLBI	National Heart, Lung, and Blood Institute
NIAID	National Institute of Allergy and Infectious Diseases
NIDA	National Institute on Drug Abuse
NIDDK	National Institute of Diabetes and Digestive and Kidney Diseases
NIGMS	National Institute of General Medical Sciences
NIH	National Institutes of Health
NIMH	National Institute of Mental Health
NINDS	National Institute of Neurological Disorders and Stroke
NIOSH	National Institute for Occupational Safety and Health
NPG	Nature Publishing Group
OECD	Organisation for Economic Co-operation and Development
OHRP	Office for Human Research Protections (formerly OPRR)
OIG	Office of the Inspector General

OPRR	Office for Protection from Research Risks (currently OHRP)
OSTP	Office of Science and Technology Policy
PHGU	Public Health Genetics Unit
PHS	Public Health Service
PIN	(Health Care) Personal Information Nondisclosure (Act)
RFP	Request for proposals
SACGT	Secretary's Advisory Committee on Genetic Testing
STD	Sexually transmitted disease
S&TPI	Science and Technology Policy Institute
TB	Tuberculosis
UNAIDS	Joint United Nations Programme on HIV/AIDS
UNESCO	United Nations Educational, Scientific and Cultural Organization
UK	United Kingdom
U.S.	United States
VHA	Veterans Health Administration
WHO	World Health Organization

Chapter One

INTRODUCTION

BACKGROUND ON THE NATIONAL BIOETHICS ADVISORY COMMISSION

The National Bioethics Advisory Commission (NBAC) was established by Executive Order 12975 in October 1995. The purpose of the commission was to advise and make recommendations to the National Science and Technology Council, chaired by the president of the United States, and to federal agencies and other entities on "bioethical issues arising from research on human biology and behavior, and the applications, including the clinical applications, of that research." NBAC was established in response to: (1) proposals by the National Institutes of Health (NIH), the Department of Energy, and other research-oriented agencies; (2) recommendations of the Advisory Committee on Human Radiation Experiments (ACHRE); and (3) a recognized need by the White House Office of Science and Technology Policy (OSTP) for a national commission to address a broad set of ethical issues, including genetic privacy and the protection of human research participants. NBAC was the fifth federal bioethics commission created to contribute to the development of public policy and to promote a national discussion on complex bioethical issues.[1]

---

[1]The four other federal bioethics commissions were the National Commission for the Protection of Human Subjects of Biomedical and Behavioral Research (1974–1978), the Ethics Advisory Board (1978–1980), the President's Commission for the Study of Ethical Problems in Medicine and Biomedical and Behavioral Research (1980–1983), and the Biomedical Ethics Advisory Committee (1988–1989).

The NBAC charter was signed on July 26, 1996, by John H. Gibbons, then assistant to the president for Science and Technology Policy. The charter specified that NBAC's first priority was to consider the "protection of the rights and welfare of human research subjects; and issues in the management and use of genetic information." President Clinton appointed the NBAC commissioners in the summer of 1996, and NBAC met for the first time on October 4, 1996. The original executive order, which expired on October 3, 1997, was amended by the president on May 16, 1997, to extend NBAC's charter until October 3, 1999. On September 16, 1999, the president extended NBAC's term for two additional years, via Executive Order 13137. On October 20, 1999, a new charter was signed by Department of Health and Human Services Secretary Donna Shalala. Executive Order 13137 was not renewed and NBAC was terminated on October 3, 2001.

THE NBAC REPORTS

During its tenure, NBAC met 48 times and submitted six major reports to the White House, three of which were requested by the president and three of which were initiated by NBAC.[2] These six reports offered 120 recommendations designed to improve the protection of human participants in research while supporting the continued advancement of science and promoting ethically sound research. Brief summaries of the reports are provided here, and more detailed information about each report can be found in the following chapters.

- *Cloning Human Beings* (June 1997). The announcement of the birth of Dolly, the cloned sheep, prompted President Clinton to urgently request advice from NBAC on the legal and ethical issues raised by cloning and its potential use in human beings. In response, NBAC concluded that the creation of "a child" by somatic cell nuclear transfer[3] was scientifically and ethically objec-

---

[2]As of this writing, NBAC reports are available at www.georgetown.edu/research/nrcbl/nbac/pubs.html or through the U.S. Department of Commerce, Technology Administration, National Technical Information Service, Springfield, VA 22161, 1-800-553-6847.

[3]*Somatic cell nuclear transfer* is a cloning technique, which involves transplanting the genetic material from a somatic cell (any cell in the body except reproductive cells) into an egg, from which the nucleus (the cellular organelle that contains the genetic material) has been removed to produce a genetically identical individual.

tionable because "current scientific information indicates that this technique is not safe to use in humans at this point." NBAC went on to say, "Moreover, in addition to safety concerns, many other serious ethical concerns have been identified, which require much more widespread and careful public deliberation before this technology may be used." NBAC recommended that the existing moratorium on attempts to create a child through cloning be continued and that the president immediately ask for voluntary compliance by the private sector. NBAC also recommended that federal legislation be enacted "to prohibit anyone from attempting, whether in a research or clinical setting, to create a child through somatic cell nuclear transfer cloning."

- *Research Involving Persons with Mental Disorders That May Affect Decisionmaking Capacity* (December 1998). A controversial history of research involving persons with mental disorders, and priorities established in Executive Order 12975, which directed NBAC to consider the protection of human research participants, prompted NBAC to consider how ethically acceptable research could be conducted with persons who suffer from mental disorders that may affect their decisionmaking capacity. In this report, NBAC recommended additional regulations to ensure that such persons are appropriately protected from harm. Other recommendations in the report call for specific action by government agencies, institutions, organizations, institutional review boards (IRBs),[4] investigators, and others responsible for the protection of research participants.

- *Research Involving Human Biological Materials: Ethical Issues and Policy Guidance* (August 1999). Concerns about protecting research participants from harm during research using those participants' biological materials, and a mandate in Executive Order 12975 to consider "issues in the management and use of genetic information," led NBAC to consider the ethical, legal, and social issues associated with the research use of human biologi-

---

[4]An *institutional review board* is a panel of physicians, scientists, and laypeople assembled by the NIH or by a company, university, hospital, or any other entity conducting clinical trials. IRBs review research plans to ensure that research participants are provided with adequate informed consent and are not exposed to unreasonable risks. IRBs also conduct continuing review of approved research to ensure that human-subject protections remain in force.

cal materials. In this report, NBAC recommended that current federal regulations be clarified to ensure that human biological materials currently stored in various repositories throughout the country and materials yet to be collected are used in an ethically appropriate manner. The report also makes recommendations about improving the processes by which informed consent should be obtained when such materials are collected in the future, and provides guidance to investigators and IRBs regarding the use of human biological materials in research.

* *Ethical Issues in Human Stem Cell Research* (September 1999). The announcement of the discovery of human embryonic stem cells[5] prompted President Clinton to turn to NBAC for advice on the ethical issues associated with research involving human stem cells. In this report, NBAC focused primarily on the president's request for recommendations outlining the conditions under which such research could be eligible for federal funding. NBAC recommended that federal sponsorship of research that involves the derivation and use of human embryonic stem cells and human embryonic germ cells[6] should be limited in two ways: (1) research should be limited to using only embryos remaining after infertility treatments or cadaveric fetal tissue, and (2) sponsorship of research should be contingent on an appropriate and open system of national oversight and review.

* *Ethical and Policy Issues in International Research: Clinical Trials in Developing Countries* (April 2001). Several issues prompted NBAC to address the topic of international research, including controversy over certain clinical trials in developing countries, in particular trials that were testing drugs to reduce mother-to-infant transmission of human immunodeficiency virus (HIV); the changing landscape of international research; and suggestions by the public that NBAC's mandate to examine the protection of human research participants extended to U.S.-sponsored or

---

[5]*Human embryonic stem cells* are cells derived from the inner cell mass of a six- to seven-day-old embryo. These cells have the ability to divide indefinitely and to give rise to all of the cell types of the adult body.

[6]*Human embryonic germ cells* are cells derived from the primordial reproductive cells of the developing fetus; they have properties similar to those of human embryonic stem cells.

U.S.-conducted international research. In this report, NBAC recommended that clinical trials subject to U.S. regulation that are conducted or sponsored in developing countries should be directly relevant to the health needs of the population of the host country. Other recommendations addressed the choice of research designs, the informed consent process, post-trial access to successful research products, and ethics review of research. NBAC also recommended that the Food and Drug Administration (FDA) should not accept data obtained from clinical trials in developing countries that do not follow the substantive ethical procedures outlined in the NBAC report.

* *Ethical and Policy Issues in Research Involving Human Participants* (August 2001). NBAC prepared its final report in accordance with Executive Order 12975, which states that "NBAC shall direct its attention to consideration of protection of the rights and welfare of human subjects," and in response to a request from the White House to undertake a thorough examination of the federal system of oversight for the protection for human research participants. In this report, NBAC concluded that the federal oversight system should protect the rights and welfare of all human research participants, regardless of whether the research is publicly or privately funded. NBAC recommended that there be a unified, comprehensive federal policy embodied in a single set of regulations and guidance, and called for legislation to create a single, independent federal office responsible for the oversight of human research participants.

RESPONSE TO NBAC'S REPORTS AND RECOMMENDATIONS

NBAC's reports and recommendations have informed the United States and foreign governments, international groups, the research community, and the public about the various dimensions of complex bioethical issues and their importance in public policy development. This report describes the response[7] of the president, Congress, fed-

---

[7]The term "response" is just one of many possible words that could be used to describe the reaction of those who have taken notice of NBAC's work and have produced documents that refer to or are based on NBAC's reports and recommendations. Al-

eral and state government, professional societies, philanthropic foundations, organizations representing private industry, patient advocacy groups, other countries, and international organizations to each of NBAC's six reports. These entities are responsible for decisionmaking and policy formulation in this country and abroad. Therefore, their response to NBAC's reports and recommendations is an indication of the extent to which NBAC's work is reflected in public policy and discourse.

However, translating the response to NBAC's work into a measure of NBAC's impact on public policy is not an easy task, for many reasons:

- First, the impact will have both a short-term and a long-term effect, and only the former can be addressed in this report.

- Second, NBAC's earlier reports have a greater likelihood of being considered in policy discussions than those released late in NBAC's tenure because interested parties have had a greater amount of time to respond to the earlier reports.

- Third, the issues addressed by NBAC differed in their immediate importance to policymakers. For example, cloning and stem cell research were of immediate concern to policymakers and received the greatest attention, whereas clinical trials research in developing countries received far less attention.

- Fourth, when a government agency, professional society, or international organization develops policy or adopts guidelines that make use of or refer to NBAC's work, it does not necessarily mean that NBAC should be credited with having had a direct influence on the decision to develop or adopt guidelines. Many factors influence the development of policy and guidelines, and a group's report may be mentioned because it truly was influential, or it may be mentioned for purely political reasons.

- Finally, much of NBAC's contribution to the policymaking process has been to inform public discussion and debate over some

---

though the word "response" could be interpreted to imply a cause-and-effect relationship, it is used more generally in this report to reflect the many ways in which NBAC's work was mentioned or used by various entities (e.g., government, societies, international organizations).

highly contentious and value-laden issues, a contribution that is not easily measured.

Acceptance of NBAC's recommendations is not, therefore, a full measure of the commission's contribution to public policy. It is important that any analysis of NBAC's contribution to public policy be conducted with these limitations in mind.

The various entities responded to NBAC's reports and recommendations in a number of ways.[8] The president, Congress, federal agencies, professional societies, the private sector, and others have issued statements, guidance, and other documents that referenced, discussed, or were based on NBAC's reports and recommendations. In addition, legislation was introduced at both the federal and state level that mentioned NBAC or was based on NBAC's recommendations. The following responses to NBAC's work were tracked and assessed by RAND:

- Responses by the president of the United States and other members of the administration to NBAC's reports and the recommendations in them

- Federal and state legislation introduced or enacted that was a result (direct or indirect) of NBAC's reports and recommendations

- Testimony before Congress by NBAC commissioners, staff, and others at Congress's invitation

- Responses to NBAC's reports and recommendations, including guidelines, statements, policies, and procedures, by relevant federal agencies

- Statements and related documents from professional societies, private industry, and other entities that followed from or were made in response to NBAC's reports and recommendations.

Other countries and international organizations have developed policies or issued statements on topics on which NBAC has reported.

---

[8]The appendix provides a detailed explanation of the methods used to track the response to NBAC's reports and recommendations. Detailed information about the documents that were used and how they were collected is also provided in the appendix.

Many of these policies and statements cite the commission's work. NBAC commissioners and staff were also consulted by international organizations such as the World Health Organization (WHO) and the United Nations Educational, Scientific and Cultural Organization (UNESCO). The following responses to NBAC's reports and recommendations by other countries and international organizations were tracked and assessed:

- Legislation that was introduced or enacted abroad that was a result (direct or indirect) of NBAC's reports and recommendations

- Policies, statements, and other documents from national scientific, medical, and bioethics bodies and international organizations that refer to or are based on any of NBAC's reports and recommendations.

NBAC's work has been discussed in leading national newspapers and magazines and on television and radio. In addition, several scholars have written about NBAC and its reports and recommendations in prominent bioethics, medical, and scientific journals. Therefore, this report includes a discussion of the media reports and articles in the academic literature that describe NBAC's work or discuss its reports and recommendations (see Chapter Two).

Government Response

NBAC's primary function, as established in Executive Order 12975, was to provide advice to the federal government on matters pertaining to bioethics. In its reports, NBAC made several recommendations to Congress and federal agencies that called for a variety of actions by these entities, including legislation, regulation, and additional guidance, and support for both professional and public education. In addition, NBAC recommended that the states also enact legislation.

NBAC's reports prompted responses from the White House, Congress, state governments, and federal agencies. In addition, NBAC commissioners and staff were invited to testify on several NBAC reports before the House and the Senate. Presidential documents, federal and state legislation, congressional testimony, federal regulations, guidance, policy statements, and other relevant documents will be described in more detail in the following chapters.

Response from Professional Societies, Organizations, and Foundations

Many scientists and physicians belong to professional and academic societies, associations, or organizations, which often have codes of ethics governing their members' general behavior and can set voluntary, informal standards for professional behavior. In addition, philanthropic foundations and organizations, associations and organizations representing private industry, and patient advocacy groups also develop policy positions and codes of conduct.

In its reports, NBAC made various recommendations to the scientific and medical communities. A number of these recommendations were directed specifically to professional societies, funding institutions, and the private sector. In addition, NBAC also appealed to the private sector for voluntary compliance with many of its recommendations. Several societies, organizations, and foundations have looked to NBAC's work to inform their discussions and policy decisions, and have issued documents and policy statements that reference, discuss, or are based on NBAC's reports and recommendations. Many societies, organizations, and foundations also responded to NBAC's requests for public comment on draft reports, posting their remarks on their Web sites for their members and others to consider.[9] In addition to documents and policy statements that mention NBAC, several professional societies, organizations, and

---

[9]Responses to NBAC's requests for public comment on its draft reports often represent the views and policy positions of the responding societies, organizations, and foundations. In addition, comments that are posted to publicly accessible Web sites are available to both the members of these societies, organizations, and foundations and to the public at large. Therefore, public comments on NBAC reports could be considered to be responses to NBAC's work. However, highlighting some public comments solely because they are posted on a Web site may leave the reader with the mistaken impression that these comments were more important or more influential than other comments that were not publicly available. Therefore, comments provided to NBAC during the public comment periods are not included in this report. (For lists of respondents to NBAC's requests for public comment on its draft reports, see the following NBAC reports: *Research Involving Persons with Mental Disorders That May Affect Decisionmaking Capacity; Research Involving Human Biological Materials: Ethical Issues and Policy Guidance; Ethical and Policy Issues in International Research: Clinical Trials in Developing Countries;* and *Ethical and Policy Issues in Research Involving Human Participants.*)

foundations have links to NBAC's home page on their Web sites.[10] The interface between NBAC and professional groups, philanthropic foundations, organizations representing private industry, and patient advocacy groups will be described in more detail in the following chapters.

International Response

Due to the transnational nature of scientific work, a need exists for international cooperation regarding the conduct of scientific and medical research. NBAC recognized the increasingly global nature of biomedical science and the importance of international cooperation, and took this factor into consideration when it prepared its reports and recommendations. NBAC examined documents, policy statements, regulations, and legislation from other countries and international organizations to inform their deliberations on all of the topics it considered. In addition, in some of its reports NBAC recommended international cooperation. Likewise, other countries and international organizations have looked to NBAC's work to inform their discussions and policy decisions. Details about NBAC's role in the international debate on the ethical conduct of biomedical research is described in more detail in the following chapters.

ORGANIZATION OF THIS REPORT

Chapter Two describes the news coverage and articles in the academic literature that have discussed NBAC and its reports and recommendations.

Chapters Three through Eight deal with the response to each of NBAC's six reports by governmental bodies, professional societies, organizations, foundations, other countries, and international

---

[10]A comprehensive listing of societies, organizations, and foundations whose response consisted solely of links on their Web sites to the NBAC home page and/or NBAC reports is not provided in this report. While it could be argued that an organization's providing a link to the NBAC site and/or its reports represents yet another way NBAC has an impact on public policy, the true significance of providing a link to NBAC's Web site cannot be assessed. Therefore, this report focuses on statements, guidance, and other documents that directly reference, discuss, or are based on NBAC's work.

groups. These six chapters are presented in the order of the publication date of each NBAC report. Each of these chapters is divided into three sections: the government response; statements from societies, organizations, and foundations; and the international response.

Chapter Nine summarizes the findings of this study and discusses the impact that NBAC's reports and recommendations have had on policy or practice. The chapter also includes a summary of the types of policy changes that NBAC recommended (e.g., legislation, regulation, guidance, education); to whom those recommendations were directed (e.g., Congress, federal agencies, professional societies, the private sector); and the types of policies that have been adopted in response to the commission's recommendations (e.g., introduction of legislation in Congress and state legislatures, issuance of guidance and policy statements).

The appendix provides a detailed explanation of the methods used to track the responses to NBAC's reports and recommendations and information about the documents that were collected and analyzed for this report.

NBAC IN THE PUBLIC EYE

One aspect of the National Bioethics Advisory Commission's role in public policy was to foster public discussion of important bioethical issues. In several of its reports, NBAC discussed the importance of ongoing public discussion and education on the ethical and social implications of biomedical research. One way in which the public becomes informed about new developments in science and medicine and the complex bioethical issues raised by these discoveries is through reports in the media and academic literature.

An in-depth analysis of the discussions about NBAC's work that appeared in the media and in the academic literature, and the role these discussions have played in educating the public about ethical issues in biomedical research, is beyond the scope of this report. However, short descriptions of the type of coverage on NBAC that has appeared in the media and academic literature are provided in this chapter.

MEDIA REPORTS THAT MENTION NBAC

NBAC's recommendations and perspectives have been featured in many leading national publications, including the *Los Angeles Times*, *The New York Times*, *The Wall Street Journal*, and the *Washington Post*. Media reports that mention NBAC include news stories, editorials, and other accounts referring to NBAC reports and recommendations that have appeared in newspapers and magazines and on television and radio broadcasts. References to NBAC in the media range from a brief mention to an in-depth discussion of the commission's work.

A search of the Nexis news database from January 1996 through June 2001 found 1,287 media reports that mentioned NBAC (see Table 2.1). Most of these reports focused on subjects about which NBAC had prepared reports. However, several media reports mentioned NBAC or an NBAC commissioner in relation to other areas of biomedical science that raise ethical issues, such as assisted reproductive technologies, genetic testing, and gene transfer technology.

The issue that drew the most media attention to NBAC's work was the possibility of cloning human beings. Between February 1997, when the cloning of Dolly the sheep was announced, and June 2001, there were 665 media reports on cloning that mentioned NBAC (see Table 2.1). In 1997 alone, the year that NBAC issued its report *Cloning Human Beings*, there were 458 articles. Interestingly, a year later there were still 167 media reports on cloning that contained some reference to NBAC. While the number of articles that mention NBAC's work on cloning dropped over time, it is still interesting to note that the role of NBAC in the policy debate over cloning is still recognized several years after the release of the NBAC report on cloning human beings.

Several of the media reports on cloning simply mentioned that NBAC is a presidential commission that has deliberated on the issue of cloning and has written a report on the subject. Other articles provided a more in-depth discussion of the NBAC report on cloning. For example, several articles included criticisms from bioethicists and other experts of NBAC's recommendations.

By comparison, the NBAC report *Research Involving Persons with Mental Disorders That May Affect Decisionmaking Capacity* (1998) received limited media coverage, with the majority of the 33 media reports appearing in 1999 and 2000 (see Table 2.1). Many of the media reports mentioned NBAC's work in the context of controversial research done with the mentally ill. Other reports addressed NBAC's recommendations about informed consent and the need for review of studies designed to stimulate behavioral or physiological manifestations of the disease under study, studies designed to withdraw research participants rapidly from therapies, or studies that use placebo controls.

Table 2.1

Number of NBAC Citations in Nexis, by Year and by Topic

	Number of Articles[a]							
	Cloning[b]	Capacity[c]	Biological Materials[d]	Stem Cell[e]	International[f]	Oversight[g]	Other[h]	Total
1996	0	0	0	0	2	6	9	17
1997	458	1	2	1	7	23	27	519
1998	167	16	3	13	2	18	32	251
1999	23	15	16	228	0	44	50	376
2000	8	1	6	25	6	14	31	91
2001[i]	9	0	1	9	7	3	4	33
Total	665	33	28	276	24	108	153	1287

[a] From a search of Nexis for "National Bioethics Advisory Commission" and "NBAC."

[b] Citations of *Cloning Human Beings*.

[c] Citations of *Research Involving Persons with Mental Disorders That May Affect Decisionmaking Capacity*.

[d] Citations of *Research Involving Human Biological Materials: Ethical Issues and Policy Guidance*.

[e] Citations of *Ethical Issues in Human Stem Cell Research*.

[f] Citations of *Ethical and Policy Issues in International Research: Clinical Trials in Developing Countries*.

[g] Citations of *Ethical and Policy Issues in Research Involving Human Participants*.

[h] *Other* includes mentions of NBAC in articles on assisted reproductive technology, genetic research, genetic testing, gene therapy, medical ethics, and other topics.

[i] Data for 2001 represent reports in the media that mention NBAC from January 1, 2001, through June 30, 2001.

Media coverage of the NBAC report *Research Involving Human Biological Materials: Ethical Issues and Policy Guidance* (1999) was limited to 28 reports from 1996 through 2001 (Table 2.1). The majority of media reports appeared in 1999 and focused on the conflict between protecting one's privacy and promoting research efforts, and on the release of the report.

NBAC's deliberations on research with human embryonic stem cells and its recommendations in this area, as presented in *Ethical Issues in Human Stem Cell Research* (1999), also generated considerable media attention. A search of the Nexis database found 276 media reports on stem cell research that mention NBAC (see Table 2.1). While the isolation of human embryonic stem cells was first reported in November 1998, the vast majority of media reports mentioning NBAC appeared in 1999, the year NBAC issued its report on stem cell research. It is interesting to note that even though the issue of stem cell research continues to be front-page news, the mention of NBAC greatly decreased after the NIH released its final "Guidelines for Stem Cell Research" in August 2000 (National Institutes of Health, 2000).

Several of the media reports on stem cell research focus on NBAC's recommendations for federal funding of both the derivation (i.e., the production of stem cells from embryos) and use of human embryonic stem cells. Most of these media reports also compare NBAC's recommendations with those of the NIH and the American Association for the Advancement of Science (AAAS), both of which support the use of federal funds for research using human embryonic stem cells but not for the derivation of human embryonic stem cells.

The media coverage of NBAC's deliberations over the recommendations in its *Ethical and Policy Issues in International Research: Clinical Trials in Developing Countries* report and the coverage on the report itself were limited, with 24 media reports from 1996 through 2001 (see Table 2.1). NBAC was mentioned several times when the controversy over Human Immunodeficiency Virus/Acquired Immune Deficiency Syndrome (HIV/AIDS) research was first reported in 1997, and then again when NBAC was deliberating on its recommendations for its *Ethical and Policy Issues in International Research: Clinical Trials in Developing Countries* report, and once again after it issued its report in April 2001. Many of these later media reports discussed NBAC's recommendations on clinical trials that use placebo controls and on post-trial access to treatment.

Media reports mentioning NBAC that were on the topic of oversight of research involving human participants appeared as early as 1996 (see Table 2.1). Much of this early media coverage mentioned NBAC's charge to review U.S. federal protections pertaining to human research participants. In 1999, there were numerous media reports mentioning NBAC's work that were about suspensions of federally funded research at several prominent research institutions by the Office for Protection from Research Risks (OPRR; now the Office for Human Research Protections, OHRP). In 2000 and 2001, media coverage of NBAC's deliberations over the recommendations in its *Ethical and Policy Issues in Research Involving Human Participants* report and coverage of the report itself focused mainly on NBAC's recommendation for a single, independent federal office responsible for the oversight of human research participants in both the public and private sector.

PUBLICATIONS ABOUT NBAC IN THE ACADEMIC LITERATURE

Several scholars have written about NBAC and its reports and recommendations in prominent bioethics journals, including the *Hastings Center Report* and the *Kennedy Institute of Ethics Journal*, and in medical and scientific journals, such as the *Journal of the American Medical Association*, the *New England Journal of Medicine*, *Nature*, and *Science*. Citations to NBAC and discussions of its reports in the academic literature were found through searches of the on-line databases PubMed, Bioethicsline, Social Sciences Citation Index, and Science Citation Index. Articles were also found through searches of the on-line versions of the *Journal of the American Medical Association*, *New England Journal of Medicine*, *Science*, and *Nature* and other Nature Publishing Group journals.[1]

---

[1]Nature Publishing Group (NPG) is the scientific publishing arm of Macmillan Publishers Ltd. As of this writing, NPG's other publications include *Nature Biotechnology, Nature Cell Biology, Nature Genetics, Nature Immunology, Nature Medicine, Nature Neuroscience, Nature Structural Biology, Nature Reviews Cancer, Nature Reviews Genetics, Nature Reviews Immunology, Nature Reviews Molecular Cell Biology, Nature Reviews Neuroscience, Nature Reviews Drug Discovery, British Dental Journal, British*

Table 2.2 lists the number of articles citing NBAC, by year, in the on-line databases and journals, and Table 2.3 shows the number of articles that specifically discuss the NBAC reports (see the Bibliography under "Academic Literature Citing NBAC" for a list of the articles in the academic literature that discussed the commission). Articles that discussed NBAC's work were most abundant in 1999, and there was at least one article on each of NBAC's six reports, some of which were completed at the time, some of which were in progress, and some of which were just beginning. The NBAC report discussed most often in the academic literature was *Cloning Human Beings.* The next most frequently discussed report was *Research Involving Persons with Mental Disorders That May Affect Decisionmaking Capacity,* closely followed by *Ethical Issues in Human Stem Cell Research.*

Publications in the academic literature that refer to the NBAC report *Cloning Human Beings* range from those that praised the report as a balanced response to a complex bioethical issue, to those that criticized it for ducking the difficult ethical questions about human cloning by relying on the safety concerns surrounding current cloning techniques to justify its recommendations. Other publications commended NBAC for recommending a ban on the use of somatic cell nuclear transfer to create "a child" while protecting freedom of scientific inquiry and allowing important research activities to proceed. And still others argue that the proposed ban on somatic cell nuclear transfer would infringe upon scientific freedom of inquiry and procreative liberty interests.

Research Involving Persons with Mental Disorders That May Affect Decisionmaking Capacity has been criticized in the academic literature by mental health advocates who believe the protections it recommends do not go far enough and by psychiatrists who believe the report's recommendations would impede psychiatric research. Recurring criticisms of the report were that (1) the scope of the report was too narrow and would stigmatize individuals with mental disorders, (2) NBAC's recommended guidelines for assessing decisionmaking capacity were too burdensome, and (3) NBAC's categories of research risk were either too restrictive or too loose.

Journal of Cancer, Nature Science Update, Cancer Update, Physics Portal, and *Materials Update.*

Table 2.2

Number of Articles in the Academic Literature Citing
NBAC, by Year

Year	Number of Articles
1996	2
1997	17
1998	12
1999	30
2000	17
2001[a]	7
Total	82

[a]Through July 2001.

Table 2.3

Number of Articles in the Academic Literature Citing Each NBAC Report

NBAC Report	Number of Articles
Cloning Human Beings	30
Research Involving Persons with Mental Disorders That May Affect Decisionmaking Capacity	18
Research Involving Human Biological Materials	11
Ethical Issues in Human Stem Cell Research	17
Ethical and Policy Issues in International Research: Clinical Trials in Developing Countries	4
Ethical and Policy Issues in Research Involving Human Participants	4

NOTE: Articles that discuss more than one of NBAC's reports are accounted for under each report listed. Articles that mention NBAC, but do not specifically discuss any of the NBAC reports, are not listed in this table.

Research Involving Human Biological Materials: Ethical Issues and Policy Guidance was described in the academic literature as containing the core elements of acceptable policy and a reasonable balance between safeguarding privacy and advancing the needs of research. Other articles discussed NBAC's recommendations on informed consent for research involving human biological materials. A working draft of the *Research Involving Human Biological Materials* report received attention in the academic literature early in 1998, prompting comments from the scientific community and from NBAC itself.

References in the academic literature to the *Ethical Issues in Human Stem Cell Research* report tended to either recount the commission's recommendations or use the report in analyses of arguments for and against federal funding of embryonic stem cell research. Several of the articles agreed with NBAC's conclusion that there is no moral distinction between the derivation of and the use of human embryonic stem cells, and supported NBAC's recommendation that federal funding should be allowed for both the derivation and use of stem cells from embryos remaining after infertility treatments.

Because the *Ethical and Policy Issues in International Research: Clinical Trials in Developing Countries* and *Ethical and Policy Issues in Research Involving Human Participants* reports were not issued until April 2001 and August 2001, respectively, few articles had appeared in the academic literature about these reports as of the writing of this document. The articles that were found simply mentioned that NBAC was working on these issues. As further consideration is given to the issues addressed by these two reports, it may be worthwhile to survey the academic literature again for any discussions of these two reports.

Chapter Three

RESPONSE TO THE *CLONING HUMAN BEINGS* REPORT

On February 23, 1997, the world first learned of Dolly, the cloned sheep. The announcement about the birth of Dolly, the first mammal to be cloned from an adult cell,[1] brought into sharp focus the future possibility of cloning human beings along with all its inherent moral, ethical, and legal implications. The next day, President Clinton asked NBAC to advise him within 90 days on the legal and ethical issues that cloning raises in regard to its potential use in human beings.

NBAC issued its *Cloning Human Beings* report on June 9, 1997. In its recommendations to the president, NBAC concluded that the "creation of a child" by somatic cell nuclear transfer is scientifically and ethically objectionable at this time.[2] NBAC recommended that the existing moratorium on attempts to create a child through cloning be continued and that the president immediately ask for voluntary compliance by the private sector. NBAC also recommended that federal legislation be enacted "to prohibit anyone from attempt-

---

[1] Some mammals, including mice, monkeys, and cows, as well as sheep, had been cloned from embryonic cells.

[2] To create a child by cloning, stated simply, one would first take a cell from the person to be cloned and then transfer the genetic information from that person's cell to an enucleated egg (i.e., somatic cell nuclear transfer) thereby creating an embryo that contains the genetic information of the person being cloned. The same initial steps (i.e., somatic cell nuclear transfer of genetic material from an adult cell into an enucleated egg) would be performed to create an embryo from which stem cells could then be harvested and used to treat diseases such as diabetes and Alzheimer's. Because this process starts with the cells of the person who has the disease to be treated, the embryonic stem cells that are created have the identical genetic information as that person.

ing, whether in a research or clinical setting, to create a child through somatic cell nuclear transfer cloning."

Since June 1997, when NBAC presented its report to the president, several developments in the possible cloning of human beings have stirred the public's interest. In the forefront of these developments are those individuals who have announced their intentions to clone a human being.[3] In addition, the discovery of human embryonic stem cells in 1998 has raised the possibility of using cloning technology to create customized stem cells (i.e., stem cells derived from the same individual) for use as cell-based and tissue-based therapies that would not be rejected by the patient's immune system.

There have also been significant scientific advances in the field of somatic cell nuclear transfer, such as the cloning of mammals other than sheep (for example, the cloning of the last cow of a rare breed) and the cloning of genetically engineered animals that can make proteins in their milk for pharmaceutical applications. All of this has spurred intense media coverage as well as captured the attention of the public and policymakers. The policy responses to cloning of human beings have been intense and wide-ranging both in the United States and abroad. There have been legislative and regulatory actions taken at the state, federal, and international levels, and policy statements released by organizations nationally and internationally.

GOVERNMENT RESPONSE

The possibility of cloning human beings has been an issue that has captured the attention of two presidents and both houses of Congress. Soon after the announcement of the cloning of Dolly, President Clinton instituted a ban on the use of federal funds for the cloning of human beings. In *Cloning Human Beings*, NBAC recommended a continuation of the president's moratorium and issued an immediate request to all firms, clinicians, investigators, and professional societies in the private and non–federally funded sectors to

---

[3]Several individuals and groups have announced that they intend to clone a human being, including an Italian doctor, Severino Antinori; an American doctor from the Andrology Institute of America, Panos Savos; a group called Clone Rights United Front, founded by Randolfe Wicker; and Clonaid, a company founded by a group called the Raelians, who believe that humans are clones of extraterrestrial beings.

comply voluntarily with the intent of the federal moratorium. The commission further recommended the following:

> Federal legislation [should] be enacted to prohibit anyone from attempting, whether in a research or clinical setting, to create a child through somatic cell nuclear transfer cloning. It is critical, however, that such legislation include a sunset clause to ensure that Congress will review the issue after a specified time period (three to five years) in order to decide whether the prohibition continues to be needed. If state legislation is enacted, it should also contain such a sunset provision. Any such legislation or associated regulation also ought to require that at some point prior to the expiration of the sunset period, an appropriate oversight body evaluate and report on the current status of somatic cell nuclear transfer technology and on the ethical and social issues that its potential use to create human beings would raise in light of public understandings at that time (National Bioethics Advisory Commission, 1997).

President Clinton issued a legislative proposal, which mentioned NBAC, prohibiting the cloning of human beings. Bills introduced in both the House and the Senate also mentioned NBAC. In addition to responses from the president and Congress, NBAC's report was also mentioned by the Department of Energy and the FDA, and in legislation introduced into the New Jersey legislature. Congressional bills and testimony, regulations, policy statements, and other documents that mentioned NBAC's report on cloning are detailed next.

Response from the President of the United States

President Clinton addressed the issue of cloning human beings on several occasions before receiving the NBAC report. On March 4, 1997, the president issued a Presidential Directive to the heads of executive departments and agencies prohibiting the use of federal funds for cloning of human beings (White House, 1997). In the directive, the president mentioned that he had asked NBAC for a report on the legal and ethical issues associated with the use of cloning technology. Because (in the words of the Presidential Directive) the current restrictions on the use of federal funds for research involving human embryos "do not explicitly cover human embryos created for implantation and do not cover all Federal agencies," the president issued the memorandum "to make it absolutely clear that no Federal funds will be used for human cloning."

In a commencement address at Morgan State University in Baltimore on May 18, 1997, President Clinton spoke about preparing the nation for the 21st century. He focused on the promise of science and technology and the ethical and moral questions raised by new discoveries in science, using the cloning of Dolly as an example. He noted that he was awaiting recommendations from NBAC, which he had asked to conduct a thorough review of the legal and ethical issues raised by the cloning of Dolly, because "No President is qualified to understand all of the implications" of the cloning of human beings.

Upon receiving NBAC's report, President Clinton proposed the "Cloning Prohibition Act of 1997" on June 9, 1997. In messages to the Senate (PM 46, June 9, 1997) and the House of Representatives (H. DOC. NO. 105-97, June 10, 1997) accompanying his legislative proposal, the president stated his agreement with NBAC's conclusion that the creation of a child by somatic cell nuclear transfer cloning is morally unacceptable at this time, and transmitted his proposal to Congress to implement NBAC's recommendations.

Consistent with NBAC's recommendation, the president's legislative proposal prohibited the use of somatic cell nuclear transfer to create a human being for five years and directed NBAC to report to the president in four and a half years on whether to continue the ban. The proposal was carefully worded to ensure that it would not interfere with beneficial biomedical and agricultural activities. This legislation, therefore, would not have prohibited the use of somatic cell nuclear transfer techniques to clone deoxyribonucleic acid (DNA), cells, and tissues, and would not ban the cloning of animals. In addition, the president called upon the private sector to refrain voluntarily from using this technology to attempt to clone a human being.

A Statement of Administrative Policy from the Executive Office of the President, Office of Management and Budget, was issued on February 9, 1998, in response to Senator Trent Lott's (R-Miss.) proposed Human Cloning Prohibition Act (S. 1601) (discussed further in the next section). The statement detailed the Clinton administration's position that it did not support the passage of S. 1601 in its current form because its ban on the use of somatic cell nuclear transfer technology to create a child as well as to create embryos for research was "too far-reaching," and it would "prohibit important biomedical research aimed at preventing and treating serious and life-threatening diseases." Instead, the Clinton administration offered several amendments to S. 1601, including the following: (1) a five-year sunset on the prohibition on human somatic cell nuclear transfer technology to ensure that there is a continuing examination of the risks and benefits of the technology; (2) permitting somatic cell nuclear transfer using human cells for the purpose of developing stem cell technology; (3) striking the bill's criminal penalties and instead making any property derived from or used to commit violations of S. 1601 subject to forfeiture to the United States; and (4) striking the provision establishing a new National Commission to Promote a National Dialogue on Bioethics within the Institute of Medicine because it would duplicate NBAC's mission.

On March 16, 2001, soon after the administration of President George W. Bush took office, Dr. Harold T. Shapiro, chairman of NBAC, wrote President Bush to inform him about the NBAC's report *Cloning Human Beings* and the commission's conclusion that no research to create a human being through cloning should take place in the United States at this time. The letter was prompted by reports in the media of two groups claiming to be planning to clone human beings. In his letter, Dr. Shapiro reinforced the commission's view that cloning was not safe and would endanger not only the baby who was the clone of another individual, but probably the pregnant mother as well. Dr. Shapiro urged President Bush to "take the lead in seeking legislation that would impose a moratorium on the use of somatic cell nuclear transfer to create human beings."

Response from Congress

Legislative Action. Several bills were introduced in the 105th, 106th, and 107th Congresses that would have made it unlawful to clone a human being. Some of these bills specifically mentioned NBAC, while another proposed the establishment of a new commission to address important bioethical issues, such as cloning. As of this writing, none of these bills introduced in either the House or the Senate have been signed into law.

On February 3, 1998, the Human Cloning Prohibition Act (S. 1601) was introduced in the 105th Congress by the Senate Majority Leader Trent Lott. S. 1601 would have banned the use of somatic cell nuclear transfer technology to create a child as well as to create embryos for research purposes, and set criminal penalties for violation of the act of (1) up to ten years in prison, a fine, or both; and (2) not more than twice the amount of any gain derived from such violation.[4] S. 1601 would have also established a National Commission to Promote a National Dialogue on Bioethics within the Institute of Medicine. NBAC was mentioned during the Senate floor debate; however, Senator Lott was unable to bring the bill up for consideration in the Senate (i.e., a cloture motion to proceed to consideration of S. 1601 failed in the Senate by a vote of 42 to 54).

The Prohibition on Cloning of Human Beings Act of 1998 (S. 1602) was also introduced on February 3, 1998, by Senators Dianne Feinstein (D-Calif.) and Edward Kennedy (D-Mass.).[5] S. 1602 was based largely on the president's legislation and would have reauthorized NBAC for another ten years. NBAC's recommendations from its report *Cloning Human Beings* were listed in Section 2 of the bill under "Findings." S. 1602 was debated on the Senate floor, but was not passed by the Senate. However, the ban on federal funding that the president declared in March 1997 remained in effect.

In the 107th Congress, Senator Sam Brownback (R-Kan.) and Representative Dave Weldon (R-Fla.) introduced the Human Cloning Prohibition Act of 2001 (S. 790/H.R. 1644) on April 26, 2001. The Senate bill was referred to the Judiciary Committee, and the House bill was referred to the Judiciary and Energy and Commerce Committees. The legislation would prohibit any person from performing or attempting to perform human cloning to create a child or to create embryos for research purposes. *Human cloning* was defined in the Senate and House bills as nuclear transfer from a human somatic cell

---

[4]An identical bill, S. 1599, the Human Cloning Prohibition Act of 1998, was introduced on February 3, 1998, by Senator Christopher Bond (R-Mo.).

[5]Senator Feinstein also introduced an identical bill, S.1611, the Prohibition on Cloning of Human Beings Act of 1998, on February 4, 1998.

into an oocyte[6] whose nucleus has been removed or inactivated. The legislation called for civil and criminal penalties for human cloning. Violators would be subject to fines of up to $1 million or prison terms of up to 10 years. The bills also contained a "Sense of Congress" provision calling for the president to commission a study from NBAC or a successor group on cloning to produce human embryos solely for research. A revised version of Weldon's bill (H.R. 1644) was introduced as H.R. 2505 in July 2001; however, the NBAC study was no longer part of the revised bill. The new version was referred to the Judiciary Committee for mark-up on July 24, 2001, and was passed in an 18-to-11 vote. On August 1, 2001, H.R. 2505 passed the House of Representatives by a vote of 265 to 162. However, Senator Brownback's bill (S. 790) was not passed by the Senate.

Congressional Testimony. NBAC commissioners and staff contributed to the policy debate on cloning human beings by testifying before Congress. On several occasions, the commissioners and staff were invited to provide expert testimony on cloning and on the NBAC report on the subject:

- Soon after President Clinton requested NBAC to advise him on cloning, Commissioner Thomas Murray was invited to testify on March 5, 1997, before the Technology Subcommittee of the House Science Committee. A week later, on March 12, 1997, Commissioner Alto Charo testified before the Subcommittee on Public Health and Safety of the Senate Labor and Human Resources Committee.

- While NBAC was still drafting its report on cloning, Dr. William Raub, Deputy Assistant Secretary for Science Policy at the Department of Health and Human Services (DHHS) and then Acting Director of NBAC, testified on May 8, 1997, before the Subcommittee on Human Resources of the House Committee on Government Reform and Oversight.

- Shortly after NBAC presented its report to the president, NBAC Commissioners Harold T. Shapiro, Thomas Murray, and David

---

[6]An *oocyte* is an immature female gamete that gives rise to the ovum (egg) after meiosis. Webster's defines an oocyte is "a female gamete before undergoing changes that correspond to maturation; an egg before maturation."

Cox testified on June 12, 1997, before the Technology Subcommittee of the House Science Committee, and Commissioners James Childress and Ezekiel Emanuel testified on June 17, 1997, before the Public Health and Safety Subcommittee of the Senate Labor and Human Resources Committee.

- In response to announcements that attempts to clone a human being were imminent, NBAC Commissioner Murray testified on March 28, 2001, before the Subcommittee on Oversight and Investigations of the House Energy and Commerce Committee, and Commissioner Alexander Capron testified on June 19, 2001, before the Subcommittee on Crime of the House Judiciary Committee.

Dr. Harold Varmus, then director of NIH, mentioned NBAC's recommendations on cloning in testimony before the Subcommittee on Health and Environment of the House Committee on Commerce on February 12, 1998, in which he addressed the potential scientific and medical benefits of somatic cell nuclear transfer technology. He cited NBAC's conclusion "that there should be imposed a period of time in which no attempt [is made] to create a child using somatic cell nuclear transfer," recalling that "[s]ubsequently, over 67,000 scientists involved in reproductive biology signed a voluntary moratorium on the cloning of a human." Consistent with NBAC's recommendation about carefully written legislation that would not interfere with other important areas of scientific research, Dr. Varmus concluded that legislation "must strike a careful balance" between banning the use of cloning technology to create a human being and not impeding promising research that requires the use of cloning technology.

Response from the Department of Energy

The *Human Genome News,* sponsored by the Department of Energy's Human Genome Program, published two stories about cloning that mentioned NBAC's report on the subject. The January–June 1997 issue contained a story about the bill the president sent to Congress to outlaw the cloning of humans ("President's Bill Would Prohibit . . .," 1997). The story notes that the president was "[a]cting on a recommendation by the National Bioethics Advisory Commission." The story also mentioned NBAC's conclusion that attempts to clone a human would be premature and unacceptably risky. The story in the

January 1998 issue explained that scientists use the word "cloning" to refer to different processes, such as the cloning of DNA, cells, and animals. The story summarized NBAC's report and recommendations on cloning of human beings, noting its conclusion that attempts to clone humans are "morally unacceptable" for safety and ethical reasons ("Cloning: From DNA . . .," 1998). The story also mentioned that NBAC called for a continued moratorium on the use of federal funds for human cloning and called for legislation containing a sunset clause so that the issue could be reviewed again in three to five years.

Response from the Food and Drug Administration

At the regulatory level, the FDA continues to assert its authority over the cloning of human beings under the Public Health Service Act and the Food, Drug, and Cosmetic Act. In a "Dear Colleague" letter, dated October 26, 1998 (Nightingale, 1998), the FDA asserted that clinical research using somatic cell nuclear transfer to create a human being could proceed only after an investigational new drug application had been filed. A featured page on the FDA's Center for Biologics Evaluation and Research (CBER) Web site, entitled "Use of Cloning Technology to Clone a Human Being," reiterated the FDA's view that there are "major unresolved questions pertaining to the use of cloning technology to clone a human being which must be seriously considered and resolved before the Agency would permit such investigation to proceed." The CBER Web site also quotes from the NBAC *Cloning Human Beings* report:

> NBAC found that concerns relating to the potential psychological harms to children and effects on the moral, religious, and cultural values of society merited further reflection and deliberation (U.S. Food and Drug Administration, 2002).

In addition, Dr. Kathryn Zoon, director of CBER, referred to NBAC's conclusions that beyond the issue of safety there are unresolved issues including the broader social and ethical implications of the use of cloning technology to clone a human being in her testimony before the Subcommittee on Oversight and Investigations of the House Committee on Energy and Commerce on March 28, 2001 (Zoon,

2001a), and at a meeting of the FDA Science Board on April 13, 2001 (Zoon, 2001b).

Response from State Government (New Jersey)

On March 24, 1997, Assemblywoman Nia Gill and Assemblyman Joseph Doria, Jr., introduced Assembly Bill (A.B.) 2849 in New Jersey's 207th Legislature. A.B. 2849 would have supplemented Title 2C of the New Jersey Statutes and amended the Genetic Privacy Act, (P.L. 1996, c. 126) by making the cloning of a human being "a crime in the first degree," and establishing that "an individual's genetic information is the property of the individual." A statement at the end of the bill explained that it was intended to complement and go beyond President Clinton's March 4, 1997, Presidential Directive imposing a ban on the use of federal funds for cloning human beings and the president's request for a voluntary moratorium by the private sector, "at least until the issuance of a report on the legal and ethical implications of cloning humans by the National Bioethics Advisory Commission."

A.B. 2849 did not pass, but the same bill, which retained the statement referring to NBAC, was reintroduced on January 13, 1998, as A.B. 329 in New Jersey's 208th Legislature and again on January 11, 2000, as A.B. 289 in the 209th Legislature.

RESPONSE FROM PROFESSIONAL SOCIETIES, ORGANIZATIONS, AND FOUNDATIONS

NBAC's recommendation to continue the presidential moratorium on the use of federal funding to clone a human being was also directed to professional and scientific societies and the private sector. NBAC made the following recommendation:

> An immediate request to all firms, clinicians, investigators, and professional societies in the private and non–federally funded sectors to comply voluntarily with the intent of the federal moratorium. Professional and scientific societies should make clear that any attempt to create a child by somatic cell nuclear transfer and implantation into a woman's body would at this time be an irresponsible, unethical, and unprofessional act.

Consistent with NBAC's recommendation, several professional and scientific societies and organizations representing the biotechnology and pharmaceutical industries called for a five-year voluntary moratorium on human cloning. However, the United States Conference of Catholic Bishops was critical of the fact that the moratorium was for only five years. The American Medical Association (AMA) introduced a resolution directing the AMA to work with NBAC on cloning issues and issued two reports on cloning that refer to NBAC. NBAC was also mentioned by the American Society for Cell Biology in congressional testimony, the AAAS in two news articles, the American Society for Investigative Pathology (ASIP) in the "Public Affairs" section of its Web site, and the Pew Charitable Trusts in a paper proposing new directions for the organization. Statements, policies, and reports from societies, organizations, and foundations on cloning that mention NBAC are described in more detail next.

Response from the American Association for the Advancement of Science

Shortly after the NBAC *Cloning Human Beings* report was released it was the subject of two articles, one in the *Professional Ethics Report*, a publication of the AAAS Scientific Freedom, Responsibility, and Law Program, and one in *Science & Technology in Congress*, a newsletter of the AAAS Center for Science, Technology, and Congress ("National Call to Ban . . .," 1997, and "President's Commission Issues . . .," 1997). Both articles discussed NBAC's report and recommendations and mentioned the legislative proposal introduced by President Clinton that was based on these recommendations. Both articles pointed out that the NBAC report was narrowly focused on the use of somatic cell nuclear transfer to create a child and did not address the use of cloning to create embryos for research, thereby leaving a loophole for this type of research to proceed in the private sector.[7] Finally, both articles indicated that some members of Congress did not think that NBAC's recommendations were restrictive enough and had introduced bills that would place a permanent ban on cloning for both reproductive and research purposes.

---

[7]Because federal funding of all embryo research is prohibited, this type of research could occur only with private funds.

Response from the American Medical Association

Resolution 528, introduced at the 1997 AMA Annual Meeting, asked the AMA to "work with the National Bioethics Advisory Commission and members of Congress and the executive branch to ensure that any legislation regulating human somatic cell nuclear transfer cloning not interfere with other significant important ongoing medical research" (American Medical Association, 1999a). Resolution 528 also asked the AMA to study the scientific and ethical implications of cloning. In 1999, the AMA's Council on Scientific Affairs and the Council on Ethical and Judicial Affairs (CEJA) issued companion reports on cloning, both of which refer to NBAC's recommendations on cloning of human beings.

The June 1999 report of the AMA's Council on Scientific Affairs, (American Medical Association, 1999a) summarized the scientific basis of cloning and described the potential risks and benefits of this technology for clinical medicine and biomedical research. The Council on Scientific Affairs reviewed NBAC's *Cloning Human Beings* report, calling it a "comprehensive presentation of the arguments on all sides of these issues, based on extensive testimony from an array of experts (scientists, ethicists, theologians, legal experts) and other interested parties." The Council on Scientific Affairs recommended that the AMA support NBAC's recommendations on cloning human beings and support ongoing research and oversight of research on this issue.

The CEJA issued a companion report published in June 1999 entitled "The Ethics of Human Cloning," which addressed the ethical issues associated with human cloning (American Medical Association, 1999b). The CEJA report mentioned the NBAC recommendation for a five-year moratorium on any attempts to create a child through somatic cell nuclear transfer cloning. The CEJA recommended that physicians should not participate in human cloning at this time because of several concerns, including unknown physical and psychosocial harms and potential effects on the gene pool. An article in the AMA *American Medical News* on September 27, 1999, also mentioned NBAC's call for a five-year moratorium on the cloning of human beings, and summarized the recommendations in the CEJA report on human cloning ("Not So Fast, Dolly," 1999).

Response from the American Society for Cell Biology

On May 2, 2001, Dr. Rudolf Jaenisch testified on behalf of the American Society for Cell Biology before the Science, Technology, and Space Subcommittee of the Senate Commerce Science and Transportation Committee about the distinction between the cloning of a human being and the "therapeutic cloning" of cells to develop tissue that may ultimately be used to treat diseases (Jaenisch, 2001). In his testimony, Dr. Jaenisch stated that there was "widespread support" of NBAC's call for a voluntary moratorium on cloning for the purpose of creating a new human being. He concluded by saying that it was "premature to ban a technique that is still in the process of evolving," but if cloning legislation was needed, the American Society for Cell Biology urged that it should address only the reproduction of a human being by cloning and not interfere with the use of cloning in biomedical research to develop treatments for diseases.

Response from the American Society for Investigative Pathology

ASIP summarized NBAC's *Cloning Human Beings* report and highlighted several of NBAC's recommendations in the "Public Affairs" section of its Web site (American Society for Investigative Pathology, 1997). ASIP indicated that NBAC's report on cloning was anticipated with much interest by both investigative pathologists and the scientific community in general. ASIP described NBAC's report as being interesting and informative, stating that NBAC "prescribed a careful, considerate initial approach to an issue that will continue to evolve and attract widespread attention."

Response from the Association of American Medical Colleges

On February 3, 1998, the Association of American Medical Colleges (AAMC) called for a five-year voluntary moratorium on human cloning. AAMC noted that their call for a voluntary ban was similar to the moratorium recommended by NBAC and adopted by the Federation of American Societies for Experimental Biology (FASEB), the American Society for Cell Biology, the Pharmaceutical Research and Manufacturers Association, the Biotechnology Industry Organization

(BIO), the American Society for Reproductive Medicine, the World Medical Association, and WHO.

Dr. David Korn, senior vice president of the Division of Biomedical and Health Sciences Research of the AAMC, referred to NBAC's recommendation for a voluntary moratorium on the cloning of human beings in his testimony before the Subcommittee on Health and Environment of the House Committee on Commerce on February 12, 1998. Dr. Korn's testimony focused on AAMC's opposition to attempts to write legislation that would ban cloning research or cloning technology for fear that it would prohibit valuable research that holds great promise for understanding and treating disease. He stated that in *Cloning Human Beings*, NBAC "persuasively articulated several drawbacks to resorting to federal legislation to resolve issues of this kind." Quoting from the NBAC report, Dr. Korn stated, "A legislative ban would represent a strong obstacle to changes in policy as scientific information develops. . . . It is notoriously difficult to draft legislation at any particular moment that can serve to both exploit and govern the rapid and unpredictable advances of science."

Response from the Biotechnology Industry Organization

In a March 27, 1997, letter to President Clinton, BIO stated its support for a voluntary moratorium on the cloning of human beings while NBAC was drafting its report on cloning (Termeer and Feldbaum, 1997). BIO also sent a list of recommendations to NBAC regarding the implications of cloning technology (Biotechnology Industry Organization, 1997). BIO recommended a continuation of the voluntary moratorium on the cloning of human beings in lieu of any new federal or state law or regulation. BIO also recommended that NBAC "express its concern that a hastily drafted, poorly envisioned federal or state law could inadvertently inhibit or even deter valuable, life-saving research." BIO made several other recommendations, including that NBAC express its support for research involving the cloning of human and animal cells, the cloning of human and animal genes and proteins, the cloning of animals, and the patenting of biotechnology inventions.

In a news release issued on June 7, 1997, immediately following the release of the NBAC report on cloning of human beings, BIO stated its full support for NBAC's recommendation that the president oppose such cloning and pledged to work with NBAC, Congress, and the administration on any legislation "to ensure that it supports the continuation of responsible medical innovation" ("BIO Supports Ban on . . .," 1997). BIO reaffirmed its support for NBAC's moratorium on the cloning of human beings in a news release issued January 7, 1998, in response to news reports that a Chicago scientist planned to clone a human being ("BIO Responds . . .," 1998).

In third, fourth, and fifth editions of BIO's *Guide to Biotechnology*, BIO reiterated its support of the president's voluntary moratorium on human cloning, and stated its agreement with NBAC's conclusion that "there are grave, moral, ethical and safety concerns surrounding this issue" (Biotechnology Industry Organization, 1999c, 2000a, and 2001). BIO repeated these same assertions in *Encouraging Development of the Biotechnology Industry: A Best Practices Survey of State Efforts*" (Biotechnology Industry Organization, 2000b).

BIO was concerned that legislation could "inadvertently have a devastating impact on vital research unrelated to cloning of a human being," citing NBAC's statement that it is "notoriously difficult" to draft legislation on issues such as human cloning (Biotechnology Industry Organization, 1999a). BIO later referred to NBAC's recommendation that federal legislation to prohibit the cloning of human beings should include a sunset clause, and that the science and ethics of cloning a human being should be evaluated prior to the expiration of the sunset period (Biotechnology Industry Organization, 1999b).

BIO referred to the NBAC report on cloning of human beings in a February 1, 2001, letter to President George W. Bush, which informed him of BIO's support of the voluntary moratorium on cloning human beings and its opposition to groups who have announced plans to clone a human being (Feldbaum, 2001). In addition, on March 28, 2001, Dr. Thomas Okarma, president and chief executive officer of Geron Corporation, in testimony on behalf of BIO before the Subcommittee on Oversight and Investigations of the House Committee on Energy and Commerce, stated BIO's opposition to "human reproductive cloning" and mentioned NBAC's conclusion that cloning of human beings was "morally unacceptable" (Okarma, 2001).

Response from the Federation of American Societies for Experimental Biology

On September 18, 1997, FASEB President Ralph B. Young announced the adoption of a voluntary five-year moratorium on cloning human beings ("FASEB Endorses . . .," 1997). This announcement noted that FASEB's voluntary moratorium was "[i]n accord with the recommendations by the National Bioethics Advisory Commission" for a five-year moratorium, after which time there would be reconsideration for possible extension.

In an article on the FASEB Web site entitled "Cloning: Past, Present, and the Exciting Future," Marie Di Berardino described the potential benefits of cloning technology, including technology involving the cloning of sheep, cattle, and goats that are genetically engineered to produce large quantities of human proteins in their milk (such as insulin to treat diabetes, a clotting factor to treat hemophilia, and tissue plasminogen activator to dissolve blood clots); cloning of transgenic animals for organ and tissue transplantation; and cloning to help preserve endangered species (Di Berardino, n.d.). Di Berardino said about the NBAC report on cloning human beings and its recommendations that "[a]ny regulatory or legislative actions undertaken to effect the foregoing prohibition on creating a child by somatic cell nuclear transfer should be carefully written so as not to interfere with other important areas of scientific research."

Response from The Pew Charitable Trusts

In June 1998, the Pew Charitable Trusts issued a white paper entitled "Public Voices, Public Choices," which proposed a shift in direction of the Health and Human Services Program and described four initiatives for the Trusts to consider funding (Byrnes, 1998). In a chapter of the white paper entitled "Bioethics: Are Science and Society Keeping Pace with Each Other?" the document describes a proposed bioethics initiative. Cloning is used as an example of a new development in biomedical research that raises several ethical, legal, religious, and social implications, and NBAC's report and recommendations on cloning of human beings are summarized. The chapter also highlighted NBAC's recommendations about the importance of public education in science, and the need for widespread and continuing deliberation on the issue of cloning human beings.

In May 2001, the Pew Forum on Religion and Public Life held a "Rapid Response Event" entitled "Human Cloning: Religious Perspectives" and released an issue brief entitled "Ethics of Human Cloning" (Pew Forum on Religion and Public Life, 2001). The brief mentioned President Clinton's request to NBAC to review the ethical and legal issues associated with the cloning of humans. It also described religious perspectives on cloning and referred to a paper commissioned by NBAC entitled "Religious Perspectives on Human Cloning" written by Courtney Campbell (Campbell, 1997).

Response from the United States Conference of Catholic Bishops

The United States Conference of Catholic Bishops has referred to the NBAC report on cloning several times on its Web site (http://www.usccb.org), in its newsletters and publications in Catholic newspapers, in statements, and in congressional testimony. The group's references to the report criticized NBAC's recommendation for a five-year moratorium on the use of cloning to create a child. Its on-line publications argued that the moratorium only covers the use of cloning to produce a "live-born child," which would still allow "unlimited cloning to produce human embryos, so long as the embryos are then destroyed" ("The Complete Guide to Cloning," 2001). The Conference of Catholic Bishops feared that experiments with embryos created by cloning will allow the procedure to become perfected. The publications were also critical of the fact that the moratorium is for only five years, after which time the ban on cloning human beings could be reconsidered. The publications went on to say "[i]nstead of being a ban on cloning, the moratorium amounts to a permission slip for experimenting on embryos and a mandate for destroying them."

INTERNATIONAL RESPONSE

NBAC recognized the importance of international cooperation in the effort to prohibit the cloning of human beings. In *Cloning Human Beings*, NBAC recommended that "[t]he United States Government should cooperate with other nations and international organizations to enforce any common aspects of their policies on the cloning of human beings."

Several countries and international organizations have enacted laws or issued policy statements prohibiting the cloning of human beings. Argentina, Australia, Belgium, Canada, China, Denmark, France, Germany, Israel, Japan, Norway, Peru, Slovakia, South Korea, Spain, Sweden, Switzerland, and the United Kingdom already have laws or have announced plans to pass laws prohibiting the cloning of human beings. In addition, the Denver Summit of Eight, the Council of Europe, WHO, UNESCO's International Bioethics Committee (IBC), the European Commission, and the Human Genome Organisation (HUGO) have called for a worldwide ban on the cloning of human beings.

The Australian Academy of Science, the Australian Health Ethics Committee (AHEC), UNESCO, the British Medical Association (BMA), and the United Kingdom's Human Genetics Advisory Commission (HGAC) and Human Fertilisation and Embryology Authority (HFEA) issued statements, policies, and reports that (1) all clearly distinguish between cloning to create a human being ("reproductive cloning") and the use of cloning technology for research purposes ("therapeutic cloning"); (2) called for a prohibition on cloning to create a human being; and (3) mentioned NBAC's report. The Australian Academy of Science position statement and the HGAC and HFEA report also specifically recommended that the use of cloning technology to create embryonic stem cells should be allowed. In contrast, NBAC specifically recommended that cloning to create a human being should be prohibited, but made no recommendations on the use of cloning technology to create embryonic stem cells. National and international reports, statements, and policies on cloning that mention NBAC are described in more detail next.

Response from Australia

In February 1999, the Australian Academy of Science released a position statement entitled "On Human Cloning" (Australian Academy of Science, 1999). In its statement, the academy distinguished between "reproductive cloning" to produce a human fetus and "therapeutic cloning" to produce human stem cells, tissues and organs, and based its recommendations on this distinction. Annex 4 of the Australian Academy of Science position statement described responses from the United Kingdom and the United States on the issue of cloning techniques. The section on the United States in Annex 4 described President Clinton's moratorium on the use of federal funds for cloning human beings, his request to NBAC for advice on this issue as well as on the issue of human stem cell research, and NBAC's conclusions and recommendations on the use of somatic cell nuclear transfer technology in human beings. The position statement characterized the NBAC report on cloning as "an extensive report." It stated that while NBAC recommended a moratorium on cloning of human beings because "current scientific information indicates that this technique is not safe to use in humans at this point," it could not reach a consensus on all the ethical issues. The position statement also noted that current U.S. law restricts the use of federal funds for the use of somatic cell nuclear transfer to create an embryo solely for research purposes, whereas currently there are no federal regulations on the use of private funds for this purpose.

The Commonwealth Minister for Health and Aged Care asked the AHEC for advice on the issue of cloning of human beings. On December 18, 1998, the AHEC presented the minister with its report entitled *Scientific, Ethical and Regulatory Considerations Relevant to Cloning Human Beings* (Australian Health Ethics Committee, 1998). The AHEC report drew a distinction between the cloning of a "whole human individual" and the cloning of the component parts of a human, such as DNA and cells, and called for legislation to prohibit the cloning of human beings in "all Australian States and Territories." Chapter 5 of the AHEC report, which reviews international legislation and guidelines relevant to cloning, discussed President Clinton's moratorium on federal funding for cloning of human beings and his request to NBAC to examine the ethical, legal, and social implications of cloning. The AHEC report highlighted NBAC's recommendation for federal legislation to prohibit anyone from attempting to create a child through somatic cell nuclear transfer cloning on the grounds that it is morally unacceptable because current scientific information indictates that the technique is not safe to use in humans at this time. The AHEC further commented, "Clearly, the issue of safety rather than ethical judgment was uppermost in the formulation of the recommendation." The AHEC also mentioned that President Clinton's proposed legislation to prohibit cloning required NBAC to review the prohibition within five years of its being enacted.

The Australian House of Representatives Legal and Constitutional Affairs Committee Bulletin addressing public forums on human cloning mentioned President Clinton's request to NBAC for recommendations on cloning of human beings (Parliament of Australia, n.d.). The main focus of the forums, which began in March 2000, was to obtain public input on AHEC's *Scientific, Ethical and Regulatory Considerations Relevant to Cloning Human Beings* report. In addition to the public forums, the chair of the committee, Kevin Andrews, was briefed by Eric Meslin, then executive director of NBAC, about these issues.

Response from the United Kingdom

In "Cloning Issues in Reproduction, Science and Medicine," published in December 1998, the HGAC and HFEA recommended that somatic cell nuclear transfer to create embryonic stem cells should be allowed, while the ban on using cloning to create babies should be upheld (Human Genetics Advisory Commission and Human Fertilisation and Embryology Authority, 1998). The scientists at HGAC and HFEA advised the UK Secretary of State for Health to consider specifying in regulations two purposes for which the HFEA might issue licenses for research, so that potential benefits of this technology can clearly be explored: (1) the development of methods of therapy for mitochondrial disease and (2) the development of therapeutic treatments for diseased or damaged tissues or organs.

In its review of international developments relating to cloning, the HGAC/HFEA report mentioned NBAC's contribution to the public debate on cloning in the United States, and it pointed out that no legislation has resulted from that debate. The HGAC/HFEA report supported NBAC's recommendation instructing the U.S. government to find ways "to provide information and education to the public in the area of genetics, and on other developments in the biomedical sciences, especially where these affect important cultural practices, values and beliefs." The report also mentioned President Clinton's request to NBAC for advice on stem cell research.

"Human 'Cloning'—A Discussion Paper for the World Medical Association," prepared by the BMA discussed NBAC's *Cloning Human Beings* report (English and Nathanson, 1999). The primary aim of the discussion paper was to clarify the fundamental differences between cloning to create a human being and the use of cloning technology for research purposes, such as for basic biological research, for development of tissue for transplantation, and for preimplantation genetic diagnosis. Some of NBAC's thoughts on the motives for using somatic cell nuclear transfer and the rights of the child born as a result of an individual being cloned, which the commission presents in the *Cloning Human Beings* report, are highlighted in Part 2 of the discussion paper entitled "Techniques Intended to Bring into Existence Genetically Identical Individuals—'Reproductive Cloning.'"

The BMA concluded that there is international support for a prohibition on the use of cloning to create a human being based mainly on the issue of safety, and cited several of the statements and directives from organizations such as NBAC, the World Health Organization, the European Commission, the Council of Europe, and UNESCO. The BMA also concluded that it is less likely that there will be "universal agreement" on techniques that involve the use or creation of embryos for research purposes, such as developing compatible tissue for transplantation purposes.

Response from the United Nations Educational, Scientific and Cultural Organization

The *Universal Declaration on the Human Genome and Human Rights* of the IBC was unanimously adopted by UNESCO's 186 member states on November 11, 1997. Article 11 of the declaration addresses the issue of cloning of human beings, stating: "Practices which are contrary to human dignity, such as reproductive cloning of human beings, shall not be permitted." A Web page on the UNESCO Office of Public Information site, "Reproductive Human Cloning: Ethical Issues," described the *Universal Declaration on the Human Genome and Human Rights* and its impact on the debate on cloning of human beings and contained the UNESCO Director-General's statement on human cloning (UNESCO, 1998). The Web page also provided information about the response to human cloning in countries including Argentina, Bulgaria, Canada, China, France, Germany, the

Russian Federation, the United Kingdom, and the United States. The discussion on the United States quoted NBAC's conclusion that "at this time it is morally unacceptable for anyone in the public or private sector, whether in a research or clinical setting, to attempt to create a child using somatic cell nuclear transfer cloning."

The 29th Session of UNESCO's General Conference met in Paris from October 21 through November 12, 1997. The general conference consists of representatives from UNESCO's member states, and it determines UNESCO's policies and primary lines of work. It meets every two years and is attended by member states and associate members, together with observers for nonmember states, intergovernmental organizations, nongovernmental organizations, and others.

In Volume 2 of the *Records of the General Conference* from that session, the "Report of Commission III" chapter contains a section on "Statements Made after the Adoption of the Universal Declaration on the Human Genome and Human Rights" (UNESCO, 1997). In that section, the "Statement by the Observer of the United States of America" says that the United States would like to change the following wording from Article 11 of the *Universal Declaration on the Human Genome and Human Rights*: "Practices which are contrary to human dignity, such as reproductive cloning of human beings, shall not be permitted." The United States would have preferred that the declaration use the NBAC terminology adopted by the Group of Eight at the Denver Summit, changing the first sentence of Article 11 to read: "The use of somatic cell nuclear transfer technology to create a child is not permitted."

Chapter Four

RESPONSE TO THE *RESEARCH INVOLVING PERSONS WITH MENTAL DISORDERS THAT MAY AFFECT DECISIONMAKING CAPACITY* REPORT

Many important and clinically relevant scientific findings about the causes and treatments of mental disorders have been discovered through research. However, some research involving persons with mental disorders has been controversial, especially studies in which medication is discontinued or in which research participants are given a pharmacologic or psychologic challenge[1] to induce symptoms. Some of this research has even led to government sanctions and lawsuits against the researchers and their institutions.

Although existing federal regulations for research involving human participants provide special protections for vulnerable populations, persons with mental disorders who may have impaired capacity have not received such protections. Since the 1970s, a number of unsuccessful attempts have been made to extend greater regulatory protections to people with mental disorders who serve as research participants and whose decisionmaking ability may be limited.

NBAC recognized an unmet need in this area, and in December 1996 began work to determine how ethically acceptable research could be conducted with persons who suffer from mental disorders that may affect their decisionmaking capacity. NBAC's focus was on whether additional protections were needed for research with this vulnerable population, and, if so, what those protections should be and how they should be applied. Over the 18-month period that NBAC considered these issues, the commission received input from several

---

[1]An example of a "pharmacologic challenge" would be giving an individual a drug that induces certain symptoms; an example of a "psychologic challenge" would be placing an individual in a situation that induces certain symptoms.

sources including expert testimony, commissioned papers, and interaction with professional and patient groups. In addition, NBAC issued a draft report for public comment on July 1, 1998, and received 118 responses.

In December 1998, NBAC issued the *Research Involving Persons with Mental Disorders That May Affect Decisionmaking Capacity* report. Although the ethical treatment of human research participants is required under current U.S. regulations, NBAC noted that there was no specific guidance for IRBs and investigators regarding research that involves participants with mental disorders that may impair their decisionmaking capacity. NBAC recognized a particular need for guidance in the informed consent process and in determining who may decide whether an individual with a mental disorder can or should participate in research. NBAC concluded that this area of research should be governed by specific further regulations to ensure that research participants with mental disorders that may affect their decisionmaking capacity are appropriately protected from harm.

NBAC made 21 recommendations in this area, a number of which proposed the development of new regulations for the protection of human research participants. The recommendations addressed six general areas: review bodies; research design; informed consent and capacity; categories of research; surrogate decisionmaking; and education, research, and support. They were directed to investigators and IRBs; health professionals; state legislatures; NIH, DHHS, and other federal agencies subject to the Federal Policy for the Protection of Human Subjects (the Common Rule, codified at 45 CFR Part 46); and others responsible for the protection of human research participants.

GOVERNMENT RESPONSE

In *Research Involving Persons with Mental Disorders That May Affect Decisionmaking Capacity*, NBAC recommended the development of new federal regulations for the protection of human research participants. Other recommendations were directed to the NIH, DHHS, OPRR, and federal agencies subject to the Common Rule. Two recommendations (Recommendations 15 and 16) were directed to state legislatures.

NBAC's report prompted responses from the president of the United States, Congress, and two state legislatures. The NIH took NBAC's recommendations into consideration when it developed its guidance on research involving individuals with questionable capacity to consent. In addition, several institutes at the NIH, in particular the National Institute of Mental Health (NIMH), and the National Institute for Occupational Safety and Health (NIOSH) of the Centers for Disease Control and Prevention (CDC) released policy statements and reports that referred to the NBAC report. State and federal legislation, congressional testimony, regulations, policy statements, and other relevant documents that mention NBAC are detailed in the following subsections.

Response from the President of the United States

The *Research Involving Persons with Mental Disorders That May Affect Decisionmaking Capacity* report was completed in December 1998 and submitted to the president on January 8, 1999. Upon receipt of the report, the president asked his science advisor, Neal Lane, to ensure that all agencies that conduct research involving human participants review the report and respond to its recommendations. On February 26, 1999, the Committee on Science of the National Science and Technology Council, whose members represent the federal agencies and departments engaged in science, were asked to review the report, provide formal comments about NBAC's recommendations, and describe any plans their agencies had for either implementing the recommendations or making related policy changes. Responses were requested by May 28, 1999; however, none of the federal agencies responded to this request for input.

Response from Congress (Congressional Testimony)

Eric Meslin, then executive director of NBAC, testified before the Subcommittee on Oversight and Investigations and the Subcommittee on Health of the House Committee on Veterans Affairs on April 21, 1999. NBAC Commissioner Eric J. Cassell testified before the Subcommittee on Criminal Justice, Drug Policy, and Human Resources of the House Committee on Government Reform on December 9, 1999. In their invited testimonies, both Meslin and Cassell described NBAC's recommendations in *Research Involving Persons with Mental*

Disorders That May Affect Decisionmaking Capacity, including recommendations on review bodies, research design, informed consent and capacity, categories of research, surrogate decisionmaking, and education, research, and support. Meslin and Cassell stated that NBAC's recommendations would "both enhance existing protections and facilitate broad public support for continued research on mental disorders."

Response from the Department of Health and Human Services

Consistent with its charter, NBAC wrote to several federal departments in October 1999 to request a response within 180 days to the recommendations directed to them in *Research Involving Persons with Mental Disorders That May Affect Decisionmaking Capacity*. On January 16, 2001, Donna Shalala, then secretary of Health and Human Services, sent the DHHS response, which was prepared by a multiagency working group convened by the Office of Science Policy, Office of the Assistant Secretary for Planning and Evaluation (HHS Working Group . . ., 2001). In that response, the working group commended NBAC for "its thoughtful and insightful report" that "makes a significant contribution to critically important issues in human subject research." The group agreed with many of NBAC's concerns and recommendations, and it developed proposals for how DHHS could be responsive to those recommendations. However, the group found the scope of the NBAC report, which focused specifically on persons with mental disorders that may affect decisionmaking capacity, to be too limited, noting that "some physical disorders or conditions also may result in impaired capacity to make decisions and therefore the ability to give voluntary informed consent to research."

To avoid stigmatization of persons with mental disorders and to ensure the protection of all research participants who may have impaired decisionmaking capacity, the DHHS working group concluded that the recommendations in *Research Involving Persons with Mental Disorders That May Affect Decisionmaking Capacity* should be "applicable to all persons with decisional impairment, irrespective of diagnosis." The group was also concerned that the new regulatory framework recommended by NBAC could result in an "unnecessarily lengthy review process" and "would alter IRB authority in ways that could produce anomalous results."

Response from the Centers for Disease Control and Prevention (National Institutes for Occupational Safety and Health)

"Protecting Workers with Developmental Disabilities," a report by NIOSH, an institute of the CDC, discusses issues affecting the health and safety of workers with developmental disabilities, provides recommendations for protecting those workers, and identifies issues still needing research ("Protecting Workers . . .," 2000). The conclusions in the report highlighted the need for additional research, but cautioned that the ability of workers with developmental disabilities to give informed consent for participation in research will vary considerably and needs to be evaluated. The NBAC report *Research Involving Persons with Mental Disorders That May Affect Decisionmaking Capacity* was identified as containing recommendations concerning informed consent that "should be reviewed when a study that will involve people with developmental disabilities is proposed."

Response from the National Institutes of Health

The NIH, and in particular the NIMH, being the primary place for federally funded research on mental disorders, had numerous interactions with NBAC both during and after the preparation of the *Research Involving Persons with Mental Disorders That May Affect Decisionmaking Capacity* report. Several key interactions are detailed next.

On March 11, 1999, the NIH issued "Research Involving Individuals with Questionable Capacity to Consent: Points to Consider" to provide guidance to IRBs and clinical investigators and to increase the protection of participants in research who may have questionable capacity to consent (Office of Extramural Research, 1999).

The NIH took NBAC's recommendations into consideration when developing its "Points to Consider," which NIH published in the October 15, 1999, issue of *Biological Psychiatry* (Shore and Hyman, 1999) and posted on the its Web site. Several of the points are consistent with recommendations made in the NBAC report. For example, NIH's guidance on "IRB Membership" directly parallels NBAC's recommendations on "Review Bodies" (Recommendation 1); "Use of a Surrogate" combines guidance found in NBAC's recommendations on "Legally Authorized Representatives" (Recommendations 14 and 15); and "Use of an Advance Directive" summarizes NBAC's recommendation on "Prospective Authorization" (Recommendation 13). The "Points to Consider" also states that "protections should be proportional to the severity of capacity impairment, or to the magnitude of experimental risk, or both." This guidance parallels NBAC's recommendations on "Informed Consent and Capacity" (Recommendations 6, 7, and 8) and "Categories of Research" (Recommendations 10, 11, and 12). NIH's continued support of research addressing well-validated and practical methods to assess capacity to consent is consistent with NBAC's recommendation on "Expanding Knowledge about Capacity Assessment and Informed Consent" (Recommendation 19). The "Points to Consider" document acknowledges that it is "generally consistent with the NBAC report."

On March 31, 1999, the NIH released a program announcement on "Research on Ethical Issues in Human Studies" inviting proposals for research on assessing decisionmaking capacity in general. The program announcement was designed to "encourage empirical studies that are expected to fill many gaps in our knowledge and understanding of the complex ethical issues that arise when involving human participants in research." The NIH was seeking grant applications "to develop practical and valid methods and instruments for assessing capacity to comprehend, appreciate, and reason in a research setting." This is consistent with NBAC's recommendation on "Expanding Knowledge about Capacity Assessment and Informed Consent" (Recommendation 19), which directs NIH to sponsor research on decisionmaking capacity and informed consent.

National Institute of Mental Health. In December 1997, in response to a request by NBAC for input on the issue of questionable capacity to give informed consent, NIMH held a workshop in conjunction with five other NIH institutes and the OPRR (now the OHRP). The workshop focused on issues such as the roles and responsibilities of IRBs, surrogate decisionmaking, and conflict of interest. Participants in the workshop included representatives of professional and lay advocacy programs, IRB members, and other individuals with expertise in ethical and legal issues in research involving people with questionable decisionmaking capacity. The NIMH submitted a report on the workshop to NBAC in February 1998.

A description of NBAC's draft report on research involving persons with questionable decisionmaking capacity and NIMH's response to NBAC's request for public comment were highlighted in the September 18, 1998, *Director's Report to the National Advisory Mental Health Council* (NAMHC) (National Institute of Mental Health, 1998). NIMH's comments to NBAC expressed its concerns about assessing the risks and benefits of research, assessing the capacity to consent on a case-by-case basis, and evaluating the risk of research. The *Research Involving Persons with Mental Disorders That May Affect Decisionmaking Capacity* report and NIMH's input to NBAC during the drafting of the report were again a subject of discussion at the NAMHC, this time at the group's February 1999 meeting (National Institute of Mental Health, 1999). In addition, a motion was approved at the meeting to create a new Human Subject Research Council Workgroup (HSRCW) to review psychiatric research that utilizes symptom challenge or medication discontinuation study designs before NIMH agrees to fund them. The purpose of the HSRCW, which met for the first time in July 1999, is to ensure that the scientific research in question is of sufficient rigor to justify any potential risks to human participants. This workgroup review panel is very similar to the "Special Standing Panel" proposed in Recommendation 2 in *Research Involving Persons with Mental Disorders That May Affect Decisionmaking Capacity* to "provide careful and timely evaluation of controversial research protocols."

NBAC's report on decisionmaking capacity was also discussed by Steve Hyman, director of NIMH, at the NIMH 3rd Annual Research Roundtable on June 23, 1999. In his presentation "Importance of Clinical Research and Protecting Human Subjects," Dr. Hyman commented that while NIMH shares many of the concerns articulated by NBAC, NIMH also wants to ensure that research on mental illness continues to progress and the rights and interests of research participants are protected (Hyman, 1999). He also pointed out that even though cognitive and emotional difficulties can have an impact on general medical research, NBAC chose to focus its concern on individuals with mental illnesses.

In a commentary on the NBAC report that accompanied the NIH "Research Involving Individuals with Questionable Capacity to Consent: Points to Consider" document, NIMH commended NBAC for "[its]dedication to addressing ethical issues in clinical research involving people with impaired decisionmaking capacity" (Shore and Hyman, 1999). The commentary detailed the involvement of NIMH and NIH with NBAC during the development of *Research Involving Persons with Mental Disorders That May Affect Decisionmaking Capacity*. It also related some of the research community's concerns about the report, such as the scope of the report being too narrow and the recommendations about studies involving greater-than-minimal risk but no prospect of direct medical benefit to research participants being too restrictive. NIMH was concerned that some of NBAC's recommendations might impede important biomedical research, and in September 1998 provided NBAC with examples of research that would be slowed or halted by these recommendations following a request by NBAC commissioners. The commentary concluded by thanking NBAC for "its commitment to strengthening human subject protections."

In January 2001, the NIMH issued a guidance document entitled "Issues to Consider in Intervention Research with Persons at High Risk for Suicidality" for use by investigators who perform research involving persons who are or may become suicidal, IRBs, and Data and Safety Monitoring Boards that must review and monitor these studies (National Institute of Mental Health, 2001). The guidance document points out that research involving persons with mental disorders "is receiving additional scrutiny" and refers to NBAC's report *Research Involving Persons with Mental Disorders That May Affect Decisionmaking Capacity*. The guidance document states that while NBAC's report did not address the issue of suicidality explicitly, it did raise many concerns about the role of researchers in ensuring that individuals with mental disorders are adequately informed about the risks and benefits of research studies and the availability of alternative treatments.

National Institute on Drug Abuse. The National Advisory Council on Drug Abuse (NACDA) advises the National Institute on Drug Abuse (NIDA) on NIDA-sponsored research, including review of grant applications for funding by NIDA. In September 2000, NACDA issued guidelines for the administration of drugs to human subjects, which focus on the issues that arise in research involving administration of drugs with the potential for abuse or dependence (National Institute on Drug Abuse, 2000). These guidelines are intended to identify issues to be considered by investigators and IRBs during the development and review of research protocols involving the administration of drugs to human research participants. In the section on "Administration of Drugs to Individuals with Mental Disorders," the NACDA guidelines address persons with mental disorders that may have impaired capacity to give voluntary informed consent, and encourages investigators to review NBAC's recommendations in *Research Involving Persons with Mental Disorders That May Affect Decisionmaking Capacity* about this topic.

National Institute of Neurological Disorders and Stroke. The implications of the NBAC report on decisionmaking capacity for clinical neurology and neurosurgery research were discussed at meetings of the National Advisory Neurological Disorders and Stroke (NANDS) Council of the National Institute of Neurological Disorders and Stroke (NINDS) (National Institute of Neurological Disorders . . ., 1999). Minutes of the February 11–12, 1999, meeting reflect the NANDS council's concerns with the NBAC report, especially the lack of input from researchers and patient groups other than those representing psychiatric disorders, and the lack of specific consideration of neurological disorders such as Alzheimer's disease, vascular disease, and traumatic brain injury. The NANDS council was also concerned about recommendations regarding the categories of risk to individuals involved in research, IRB membership requirements, and capacity assessment. The NANDS council concluded that if NBAC's recommendations were implemented "several possible negative effects may result for neuroscience research," and it was suggested that the neuroscience community be alerted of these concerns at upcoming national workshops and conferences. Minutes of the May 20–21, 1999, NANDS meeting summarized the NIH's concerns with NBAC's report, such as the scope of the report, the proposed new DHHS oversight plan, capacity assessment, and research involving greater-than-minimal risk with no direct benefit to research participants.

Response from State Government

In *Research Involving Persons with Mental Disorders That May Affect Decisionmaking Capacity*, NBAC addressed two recommendations to state governments regarding legally authorized representatives and durable powers of attorney for health care. Recommendation 15 directs the states to "confirm, by statute or court decision, that: (1) an LAR [legally authorized representative] for purposes of clinical decision making may serve as an LAR for research; and (2) friends as well as relatives may serve as both clinical and research LARs if they are actively involved in the care of a person who lacks decisionmaking capacity." Recommendation 16 directs the states to "enact legislation, if necessary, to ensure that persons who choose to plan for future research participation are entitled to choose their LAR."

To encourage positive state action on Recommendations 15 and 16, NBAC sent copies of *Research Involving Persons with Mental Disorders That May Affect Decisionmaking Capacity* to state governors, health commissioners, and attorneys general. With the help of Jack Schwartz, assistant attorney general and director of Health Policy Development, Maryland Attorney General's Office, a model bill was drafted and distributed to the Maryland General Assembly. Legislation based on NBAC's recommendations was introduced in Maryland and New York.

Maryland. In Maryland, Senate Bill 307, The Decisionally Incapacitated Research Subject Protection Act, was introduced by Senator Brian Frosh on February 5, 1999, and assigned to Judicial Proceedings. The purpose of the bill was to specify "requirements for research involving individuals with a potential or actual decisional incapacity," including provisions for recruitment; informed consent and assent; capacity assessment; research advance directives; and surrogates, proxy decisionmakers, or legally authorized representatives. Bill 307 differed from NBAC's recommendations in two major ways. First, it applied generally to all research participants who are or who may become decisionally incapacitated and not just persons with mental disorders. Second, the bill referred to three levels of risk associated with an individual's being involved in research—minimal risk, a minor increase over minimal risk, and more than a minor increase over minimal risk. The bill was not passed due to an unfavorable report by Judicial Proceedings.

New York. In New York, Assembly Bill 7298, An Act to Amend the Public Health Law in Relation to Human Research, which had multiple sponsors, was introduced on March 29, 1999, and was referred to the state's Committee on Health; however, the bill was not enacted. The bill would have amended New York's current Public Health Law by adding specific provisions regarding research involving persons with mental disorders that may affect decisionmaking capacity, including requirements for Human Research Review Committee membership, determination of capacity to provide informed consent, assignment of a legally authorized representative, and execution of research advance directives.[2]

Several of the proposed amendments to the New York Public Health Law corresponded to NBAC's recommendations. For example, one amendment to the Human Research Review Committee section of the Public Health Law required that "when the human research review committee reviews human research involving subjects with mental disorders that may affect decision making capacity, fifteen percent of the committee members, but no less than one member, must be a person with such a disorder or a family member of such a person, or a representative of an advocacy organization concerned with the welfare of such persons." This provision was similar to NBAC's recommendation regarding "Institutional Review Board Membership" (Recommendation 1), which has similar requirements for IRB membership.

RESPONSE FROM PROFESSIONAL SOCIETIES, ORGANIZATIONS, AND FOUNDATIONS

NBAC recommended that professional organizations develop educational materials pertaining to ethical issues associated with research involving persons with mental disorders. Other recommendations were directed to researchers and health professionals, many of whom are members of these organizations.

---

[2]A research advance directive in this case might be instructions to investigators on a research participant's wishes if the participant were unable to relate those wishes himself.

Some organizations expressed concern that some NBAC recommendations may impede research. The Alzheimer's Association funded a grant to study the impact of NBAC's recommendations. The American Academy of Neurology (AAN) and the American Neurological Association (ANA) were so concerned about the recommendations in *Research Involving Persons with Mental Disorders That May Affect Decisionmaking Capacity* that they wrote a letter to President Clinton to suggest alternatives to what NBAC had recommended. The AAMC was concerned that the scope of the report was too narrow. Statements from these associations about the NBAC report are described in more detail in the following subsections.

Response from the Alzheimer's Association

In 2000, the Alzheimer's Association sponsored a two-year study at the University of Chicago on the "Potential Impact of NBAC's Recommendations on Dementia Research," the purpose of which was to explore the potential ethical, legal, economic, and logistical consequences of new regulations proposed by NBAC with an emphasis on "identifying aspects that might impede or prohibit research" (Alzheimer's Association, n.d.). Interviews with researchers in the field were proposed to determine how the research community perceived NBAC's recommendations. The goal of the project was to help shape public policy regarding dementia research.

Response from the American Academy of Neurology and American Neurological Association

On April 14, 1999, the presidents of the AAN and the ANA wrote a letter to President Clinton to express their "great concern" that NBAC's recommendations in *Research Involving Persons with Mental Disorders That May Affect Decisionmaking Capacity* would severely limit the ability of scientists to study neurological diseases that afflict millions of Americans, such as Alzheimer's disease, Parkinson's disease, stroke, and multiple sclerosis ("Comments to President Clinton . . .," 1999). The AAN and ANA suggested alternatives to some of NBAC's recommendations, including a three-tiered system for defining research risk instead of NBAC's two-tiered system, and an investigator-based process for the initial assessment of decisionmaking capacity instead of NBAC's independent assessment. They also called for additional public forums and input from the research community and patient advocacy organizations.

Response from the Association of American Medical Colleges

AAMC, which represents the institutions primarily responsible for the conduct of clinical research in the United States, closely monitored NBAC's deliberations on research involving persons with mental disorders that may affect their decisionmaking capacity. A May 1999 article in the *AAMC Reporter* (Coleman, 1999) described the NBAC report and expressed AAMC's misgivings about the report, including the scope of the report, and the need for three distinct levels of risk to participants in psychiatric research. The 1998 AAMC President's Memoranda Advisory on "Recent Activities Related to Human Subjects Protections" described NBAC and its work on the ethical challenges of research involving the cognitively impaired, and notified readers that NBAC was seeking comments on a draft report and directed them to NBAC's Web site[3] for a copy of it (Association of American Medical Colleges, 1998).

INTERNATIONAL RESPONSE

Although NBAC's *Research Involving Persons with Mental Disorders That May Affect Decisionmaking Capacity* report prompted responses from both federal and state governments and from some professional organizations in the United States, no mention of NBAC's work in this area was found in any of the regulations, guidance, policy statements, or reports of other countries or international organizations.

---

[3]When NBAC's charter expired, the commission's entire Web site was archived at http://www.georgetown.edu/research/nrcbl/nbac, a site hosted and maintained by the National Reference Center for Bioethics Literature at Georgetown University. All of NBAC's other written materials, such as reports, meeting transcripts, briefings, congressional testimony, and such, are also archived at Georgetown.

Chapter Five

RESPONSE TO THE *RESEARCH INVOLVING HUMAN BIOLOGICAL MATERIALS: ETHICAL ISSUES AND POLICY GUIDANCE* REPORT

In the United States, human biological materials, such as biopsy specimens obtained for diagnostic purposes, organs and tissues removed during surgery, and cells and DNA collected for research, have been used by the biomedical community for educational and research purposes for more than a hundred years. There are more than 307 million such biological materials stored in the nation's laboratories, tissue repositories, and health care institutions, with millions more gathered and stored each year (Eiseman and Haga, 1999).

These human biological materials have played a major role in the understanding and treatment of such diseases as cancer, HIV/AIDS, and heart disease. Furthermore, recent and rapid advances in biomedical research have increased the amount and kind of information obtained from biological materials, making it possible to glean clinical and genetic information about the person who provided the material from almost any minuscule sample of tissue. While many of these technological advances have been instrumental in recent biomedical discoveries, they also have raised several legal, ethical, and societal issues, including concerns about privacy and informed consent.

NBAC chose to consider the research use of human biological materials under the broad mandate of Executive Order 12975, which established NABC in October 1995, to consider the rights and welfare of human research participants and "issues in the management and use of genetic information." NBAC recognized that it was crucial that the human biological materials already in storage and those yet to be collected remain accessible under appropriate conditions and with appropriate protections for the individuals who are the sources of this material. It was also clear to NBAC that the identifiability of human biological materials (i.e., the extent to which biological material can be linked to the person from whom it was obtained) was an essential part of determining the risks and potential benefits that might occur to research participants. To assist in its deliberations, NBAC reviewed relevant scientific, ethical, religious, legal, and policy literature, commissioned scholarly papers, and invited members of the public and representatives of professional and consumer organizations to provide written and verbal testimony. In addition, NBAC issued a draft report for public comment on February 22, 1999, and received 63 comments.

After nearly two years of study, NBAC submitted its *Research Involving Human Biological Materials: Ethical Issues and Policy Guidance* report to President Clinton on July 16, 1999. In *Research Involving Human Biological Materials*, NBAC concluded that "[p]roperly interpreted and modestly modified, present federal regulations can protect subjects' rights and interests and at the same time permit well-designed research to go forward using materials already in storage as well as those newly collected by investigators and others." NBAC developed a schema to describe the type of personal information associated with samples of human biological materials as they exist in clinical facilities or other repositories and as they exist in the hands of researchers. NBAC made 23 recommendations to protect human participants from harm in research involving human biological materials by providing clarification and interpretation of federal regulations, ensuring appropriate oversight and IRB review, and providing guidance for IRBs and investigators regarding informed consent.

GOVERNMENT RESPONSE

In *Research Involving Human Biological Materials*, NBAC focused on whether existing federal regulations governing research with human participants (the Common Rule) were adequate to protect human participants from harm in research involving human biological materials. NBAC found that the Common Rule was inadequate to ensure the ethical use of human biological materials in research and required some modification. In its report, NBAC provided interpretations of several important concepts and terms in the Common Rule and recommended ways to strengthen and clarify the regulations and make their implementation more consistent. Several of NBAC's recommendations were directed to OPRR and federal agencies that are signatories to the Common Rule. The CDC, FDA, several institutes at NIH, and the Secretary's Advisory Committee on Genetic Testing (SACGT) issued guidance or reports that referred to NBAC's recommendations on research involving human biological materials. DHHS also responded to NBAC's request for a review of its report. Federal agency guidance, reports, and other relevant documents that mention NBAC are detailed later in the subsections that follow.

In addition, NBAC specifically recommended that state and federal legislators should seek to harmonize rules governing research on medical records and on human biological materials when drafting legislation (Recommendation 23). Federal legislation was introduced in both the House and the Senate on the privacy of genetic and medical information that referred to NBAC. The State of Oregon also drafted legislation to amend its genetic privacy law that directed the state's Health Division to take into account NBAC's recommendations. State and federal legislation that mentions NBAC are detailed in the subsections that follow.

Response from Congress (Legislative Action)

Legislation was introduced during the 106th Congress in both the House and the Senate on the privacy of genetic information and the protection of identifiable DNA samples. H.R. 2555, the Genetic Privacy and Nondiscrimination Act of 1999, introduced on July 19, 1999, by Representative Cliff Stearns (R-Fla.), would have required NBAC to prepare a report recommending standards for the acquisition and retention of genetic information and standards to protect identifiable DNA samples. Similar bills had been previously introduced in the 104th (S. 1416/H.R. 2690) and 105th (H.R. 341/H.R. 2198) Congresses.

In addition, legislation was introduced on the privacy of medical information—S. 578, the Health Care Personal Information Nondisclosure (PIN) Act, and its companion bill in the House, H.R. 2404, the Personal Medical Information Protection Act of 1999. This legislation directed the secretary of HHS to consider the findings of NBAC when formulating recommendations with respect to privacy of health information. The Health Care PIN Act had previously been introduced in the 105th Congress (S. 1912).

Response from the Department of Health and Human Services

On May 18, 2001, DHHS responded to NBAC's request for a review of the *Research Involving Human Biological Materials* report. The DHHS response was prepared by an interagency working group consisting of representatives from the NIH, FDA, CDC, Health Resources and Services Administration (HRSA), Indian Health Service (IHS), OHRP (formerly OPRR), Office of General Counsel, and Office of the Assistant Secretary for Planning and Evaluation. The working group commended NBAC for "its careful analysis and far-reaching recommendations," stating that the NBAC report "will undoubtedly inform DHHS policy regarding human subject protection over the next several years and beyond."

The working group concurred with the majority of NBAC's recommendations and proposed a set of actions to be implemented by the appropriate DHHS agency. For example, the group proposed that OHRP, in consultation with FDA, develop and disseminate guidance for implementation of several of NBAC's recommendations. The group also concurred with NBAC's Recommendation 6, which addresses the need for separating consent for research from consent for clinical procedures, and proposed that OHRP, in conjunction with NIH, FDA, CDC and other appropriate DHHS agencies, sponsor a workshop on informed consent.

Nevertheless, the working group found some practical problems with the implementation of some of the recommendations. For example, NBAC's recommendation pertaining to disclosure of research results to participants (Recommendation 14) appears to conflict with some provisions of the Health Insurance Portability and Accountability Act (HIPAA). HIPAA, issued after NBAC completed its report, established provisions to protect the confidentiality of personally identifiable medical information and the rights of patients and research participants to access information about themselves.

Response from the Centers for Disease Control and Prevention

CDC refers to the NBAC report *Research Involving Human Biological Materials* in a Federal Register Notice of Intent announcing DHHS's plan to revise the Clinical Laboratory Improvement Amendments (CLIA) regulations for laboratories performing human genetic testing (CDC, 2000). The notice of intent solicits comments on the recommendations of the Clinical Laboratory Improvement Advisory Committee to change current CLIA requirements to specifically recognize a genetic testing specialty. The NBAC report is referred to in Question 4 of the notice, which asks whether informed consent must be obtained by an authorized person before certain types of genetic tests are performed (e.g., screening, diagnostic, carrier, presymptomatic, or susceptibility). It was noted in the Notice of Intent that NBAC's recommendations provide guidance to research laboratories, but they do not apply to clinical interventions, quality control, or teaching. In contrast, CLIA regulations do not apply to research laboratories unless they report patient specific information.

On October 1, 1997, CDC's ad hoc Task Force on Genetics in Disease Prevention issued "Translating Advances in Human Genetics into Public Health Action: A Strategic Plan," which contained recommendations to "help ensure that results from genetic research are responsibly used in public health practice" (CDC, 1997). Even though NBAC was still in the early stages of its work on research involving human biological materials, the task force mentioned that it was aware of NBAC's work on genetic issues.

Response from the Food and Drug Administration

On July 27–28, 2000, at the Microbiology Devices Panel Meeting, Dr. Jean Toth-Allen, from the Office of Compliance, Division of Bioresearch Monitoring, announced that the FDA had adopted the same terminology as NBAC for repository collections, research samples, and identification of research participants ("Summary Minutes . . .," 2000). Based on NBAC's recommendations, the FDA is now recommending that informed consent with an IRB-approved consent form be obtained for any study submitted to FDA using prospectively collected samples and for establishing any type of tissue repository. For samples already in storage, the FDA would like to see some kind of "general consent form" that indicates that the individual knew that eventually their sample was going to be used in research. Dr. Toth-Allen warned that new privacy laws may make the requirement for informed consent even more complicated.

Response from the National Institutes of Health

National Cancer Institute. At the March 2–3, 1998, meeting of the Board of Scientific Advisors for the National Cancer Institute, the technical, ethical, social, and legal issues associated with developing informatics systems for cancer research were discussed (National Cancer Institute, 1998). During the discussion, it was noted that NBAC "will be issuing a judgment" on issues concerning informed consent for the use of human biological materials in research, "which will influence the actions of institutional review boards, OPRR, and Congress." It was also stressed that during the time that NBAC posted the *Research Involving Human Biological Materials* report on its Web site for public comment, it would be important to articulate the implications of the report for research and for patient privacy and confidentiality.

National Heart, Lung, and Blood Institute. The National Heart, Lung, and Blood Institute (NHLBI) issued its "Human Tissue Repository Guidelines" in April 2000 (National Heart, Lung, and Blood Institute, 2000). The guidelines mention that issues regarding the use of stored tissue are covered in several documents including NBAC's *Research Involving Human Biological Materials* report, OPRR's guidance document "Issues to Consider in the Research Use of Stored Data or Tissues," and NIH's draft document "Research on Human Specimens." The NHLBI guidelines state that they are consistent with these documents.

National Institute of Allergy and Infectious Diseases. In July 2000, a request for proposals (RFP) entitled "Hepatitis C Recovery Research Network" was issued by the Division of Microbiology and Infectious Diseases (DMID) of the National Institute of Allergy and Infectious Diseases (NIAID). The RFP's "Guidance for Clinical Protocol Development" section included a subsection entitled "Suggested Language for Informed Consent for Future Use of Biological Specimens Collected Under Clinical Protocols." The guidance was issued to assist researchers in meeting federal guidelines and regulations for providing informed consent to participants in clinical research. (Division of Microbiology and Infectious Diseases, 2000). It contains recommendations and an informed consent template for the storage and research use of biological specimens for purposes other than those defined in the original clinical protocol.[1] The document states that DMID's recommendations were based on current federal regulations and the recommendations in NBAC's *Research Involving Human Biological Materials* report.

National Institute of General Medical Sciences. On July 20, 1999, the National Institute of General Medical Sciences (NIGMS) held a workshop on "Population-Based Samples for the NIGMS Human Genetic Cell Repository." The purpose of the workshop was to consider the scientific benefits and the risks to certain populations from expanding the collection of population-based samples in the NIGMS Human Genetic Cell Repository and to develop recommendations about whether, and under what conditions, the repository should acquire these samples.

The NIGMS workshop report presents consensus recommendations on both scientific and human research participant issues related to the population-based samples. The NIGMS report describes *Research Involving Human Biological Materials* as providing a "thorough analysis of the ethical issues involving groups," and quotes NBAC's Recommendations 17 and 18, which address potential harms to groups (National Institute of General Medical Sciences, 1999). The NIGMS report offers additional recommendations developed from its workshop that "might be considered to afford protections beyond what were recommended in the NBAC report."

In September 2000, NIGMS convened the First Community Consultation on the Responsible Collection and Use of Samples for Genetic Research. Participants in the consultation shared their views and concerns about the conduct of genetic research with populations and communities, such as the potential for discrimination, stigmatization, and breach of privacy. The NIGMS report on the consultation commented that *Research Involving Human Biological Materials* "addresses pertinent issues and offers pragmatic recommendations for use of tissue samples" (National Institute of General Medical Sciences, 2000). Participants in the consultation noted that public in-

---

[1]The recommendations and informed consent template provide information to investigators about how to handle samples that may be used in future research that is not originally defined in a clinical protocol at the time the sample is collected.

volvement in NBAC's meetings and deliberations was essential, and they encouraged NIH to promote public involvement in the discussion of the ethical, legal, and social implications of genetic research.

Office of Human Subjects Research. In August 2000, the Office of Human Subjects Research at NIH issued "Guidance on the Research Use of Stored Samples or Data" (Office of Human Subjects Research, 2000). The definitions, policy, and implementation discussed in the document were designed to provide NIH investigators with guidance on the research use of stored samples or data, based on the NBAC *Research Involving Human Biological Materials* report. The guidance adopted NBAC's definitions for the identifiability of human biological materials and NBAC's recommendations on the need for IRB review and informed consent when using stored human samples or data.

Response from the Secretary's Advisory Committee on Genetic Testing

SACGT was chartered in 1998 to advise DHHS on the medical, scientific, ethical, legal, and social issues raised by the development and use of genetic tests. SACGT finalized its report *Enhancing the Oversight of Genetic Tests: Recommendations of the SACGT* in June 2000 (Secretary's Advisory Committee on Genetic Testing, 2000). The report was requested by Dr. David Satcher, former assistant secretary for Health and U.S. Surgeon General from 1998 to 2002. SACGT's recommendations were developed in consultation with the public,[2] informed by a literature review of more than 70 scholarly articles on genetic testing, and based on a careful consideration of the issues. The SACGT report noted that the recommendations in NBAC's *Research Involving Human Biological Materials* report were relevant to some of the issues that it addressed. SACGT concluded that additional oversight is warranted for all genetic tests, but that the oversight mechanisms should be innovative so as not to limit the development of new tests or inordinately delay their availability.

---

[2]The public was involved in several ways, including through a *Federal Register* notice, a targeted mailing to 2,500 interested individuals and organizations, a Web-based consultation, and a public meeting.

Response from State Government (Oregon)

The 1995 Oregon State Legislature passed a state law on genetic privacy establishing that "[a]n individual's genetic information is the property of the individual." The law was revised during the 1997 legislative session to read "an individual's genetic information and DNA sample are the property of the individual except when the information or sample is used in anonymous research." Senate Bill 1008, introduced during the 1999 Oregon legislative session, would have further amended the genetic privacy law to allow identifiable genetic information and identifiable DNA samples to be used for genetic research if securely coded or if the research is subject to federal regulations. This bill also directed the Oregon Health Division to define "securely coded," taking into account NBAC's recommendations. However, Senate Bill 1008 died in the Judiciary Committee.

RESPONSE FROM PROFESSIONAL SOCIETIES, ORGANIZATIONS, AND FOUNDATIONS

NBAC analyzed 14 position statements from scientific and medical organizations regarding the research use of human biological material and found little consensus over key issues, such as what constitutes an identifiable human research participant, when should informed consent be required, and what constitutes proper consent. NBAC's recommendations in the *Research Involving Human Biological Materials* report address many of these inconsistencies in the language of these statements. NBAC also recommended that NIH, professional societies, and health care organizations "continue and expand their efforts to train investigators about the ethical issues and regulations regarding research on human biological materials and to develop exemplary practices for resolving such issues."

FASEB and ASIP expressed concern about early drafts of NBAC's recommendations on research involving human biological materials. The IRB policies and procedures for the use of human biological materials in research issued by Intermountain Health Care (IHC) incorporated and referenced many of NBAC's recommendations. The Online Ethics Center for Engineering and Science provides an on-line teaching module based on NBAC's report. Statements, policies, and educational material from these societies and organizations that address the NBAC report *Research Involving Human Biological Materials* are described in more detail next.

Response from the Federation of American Societies for Experimental Biology

A story in the February 1997 *FASEB Newsletter,* a publication of the FASEB[3] Office of Public Affairs, related a concern of one of its member societies, ASIP, that NBAC's early draft recommendations on the use of human biological materials in research were "potentially far-reaching" and "bureaucratically burdensome" ("ASIP Seeks . . .," 1997). ASIP was concerned that not enough scientific input was being considered and proposed that FASEB "express its strong concern that biomedical scientists be represented on the commission." The FASEB Public Affairs Advisory Committee approved the proposal and directed the Office of Public Affairs to establish a Subcommittee on Bioethics to develop FASEB positions on emerging issues in research involving human biological materials.

Response from Intermountain Health Care

"Research Using Human Biological Materials: IRB Policies and Procedures," by Howard Mann, Chairman of the Human Subjects and Research Committee, Urban Central Region, IHC,[4] informs investigators affiliated with IHC about their obligations for meeting institutional requirements and protecting the rights and welfare of research participants (Mann, n.d.). Dr. Mann incorporated and referenced many of the recommendations in NBAC's *Research Involving Human Biological Materials* report in his document. For example, the document adopts NBAC's schema for identifiability of research samples, requirements for investigators using coded or identified samples (Recommendation 5), and criteria for waiver of consent and determination of minimal risk to research subjects (Recommendation 10).

---

[3]The Federation of American Societies for Experimental Biology is a coalition of independent member societies that represent biomedical and life scientists. FASEB plays an active role in public policy and lobbying for the interests of biomedical scientists (see http://www.faseb.org).

[4]IHC is a charitable, community-owned, nonprofit health care organization based in Salt Lake City that serves the health needs of Utah and Idaho residents.

Response from The Online Ethics Center for Engineering and Science

The Online Ethics Center for Engineering and Science at Case Western Reserve University is an on-line resource for students, faculty, postdoctoral fellows, and research staff that provides materials on its Web site (www.onlineethics.org) for understanding and addressing ethical issues that arise in science and engineering research. "Ethical Challenges in Research with Human Biological Materials," a teaching module created by NBAC Commissioner Thomas Murray, which the center published on its site, is based on the NBAC report *Research Involving Human Biological Materials* (Murray, n.d.). The module consists of background information about research with human biological materials, a categorization of repository collections and research samples of human biological materials according to their identifiability, and the identification of key issues, such as informed consent, potentially objectionable research, privacy, and commercialization. For discussion purposes, the module presents four research scenarios that raise many of the key ethical issues associated with research involving human biological materials.

INTERNATIONAL RESPONSE

NBAC examined statements addressing the ethical use of human biological materials from other countries and international organizations, and the commission found that many of the guidelines were based on ethical considerations that are common worldwide, such as respect for privacy, respect for autonomy, and the noncommercialization of human biological materials. However, NBAC also found differing positions on how to control access to and use of human biological materials. NBAC concluded that the standardization of policies regarding the use of DNA samples would facilitate international cooperation in biomedical research.

A report from the Chief Medical Officer in the United Kingdom on post-mortem removal, retention, and use of human organs and tissues and a report from the Organisation for Economic Co-operation and Development (2000) on genetic testing both refer to the NBAC report *Research Involving Human Biological Materials*. In addition, Germany's Institut für Wissenschaft und Ethik reprinted the Executive Summary of *Research Involving Human Biological Materials* in the 2000 edition of its *Yearbook for Science and Ethics,* and the Japan Pharmaceutical Manufacturers Association (JPMA) translated Volume I of the NBAC report into Japanese. National and international groups that refer to or have published or translated the NBAC report *Research Involving Human Biological Materials* are described in more detail next.

Response from the United Kingdom

A report entitled *The Removal, Retention and Use of Human Organs and Tissue from Post-Mortem Examination: Advice from the Chief Medical Officer* (Chief Medical Officer, 2001) was written in response to the recent controversy in the United Kingdom over the retention of organs and tissues removed from children at post-mortem examinations for teaching or research without the permission or knowledge of their parents. Chapter 2 of the report, "Removal, Retention, Use and Disposal of Organs and Tissues after Death: Legal and Regulatory Issues," mentions the NBAC report *Research Involving Human Biological Materials* and provides details about the collections of human biological materials in the United States that are described in the NBAC report. The Chief Medical Officer's analysis of the problems with the UK's current system led him to conclude that significant changes to current practice were needed and to recommend a new comprehensive system for the oversight of the removal, storage, and use of human tissue from the living and the dead.

Response from the Organisation for Economic Co-operation and Development

The Organisation for Economic Co-operation and Development (OECD) report *Genetic Testing: Policy Issues for the New Millennium* summarized the topics, issues, and policy considerations discussed at the OECD workshop in Vienna on February 23–25, 2000 (Organisation for Economic Co-operation and Development, 2000). The workshop was cosponsored by the European Commission and the Governments of Austria and the United Kingdom and was convened "to consider whether the approaches of OECD Member countries for dealing with new genetic tests are appropriate and mutually compatible." The workshop was attended by approximately 150 experts and delegates from the 29 OECD countries,[5] and representatives of patients' organizations, industry, nongovernmental organizations, and the World Health Organization.

The NBAC *Research Involving Human Biological Materials* report was mentioned in the OECD report in the section that addressed "security of DNA and healthcare data banking facilities." The OECD report explained that statements in the NBAC report addressing the need to develop adequate encryption technology and related mechanisms for review were similar to statements made by the HUGO Ethics Committee in "Principled Conduct of Genetic Research," the Council of Europe in its "Recommendations on the Protection of Medical Data," by the American College of Medical Genetics in "Standards and Guidelines: Clinical Genetics Laboratories," and by UNESCO. The OECD report noted that "despite the overwhelming agreement of international bodies and professional organisations alike on the need for 'appropriate technical measures' to protect data, little progress has been made in clarifying what the term 'appropriate' should signify and how this goal can be achieved in practice."

Response from Germany

Germany's Institut für Wissenschaft und Ethik published the Executive Summary of *Research Involving Human Biological Materials* in its 2000 edition of the *Jahrbuch für Wissenschaft und Ethik* (*Yearbook for Science and Ethics*). The Institut "found it very important to address the attention of the German public and scientific community to bioethical research in the USA," and indicated that it would consider publication of future NBAC reports in future editions of the journal (Hübner, 2000).

---

[5]The member countries of the OECD are Australia, Austria, Belgium, Canada, the Czech Republic, Denmark, Finland, France, Germany, Greece, Hungary, Iceland, Ireland, Italy, Japan, South Korea, Luxembourg, Mexico, the Netherlands, New Zealand, Norway, Poland, Portugal, Spain, Sweden, Switzerland, Turkey, the United Kingdom, and the United States.

Response from Japan

Volume I of *Research Involving Human Biological Materials: Ethical Issues and Policy Guidance* was translated into Japanese by JPMA, a consortium of 83 leading research-based companies (Okazaki, 2000). A 20-member research and development committee of JPMA was conducting a study on cooperation between the Japanese government and the private sector for the evaluation of the safety and efficacy of drugs using human tissues. To ensure that the committee understood the ethical problems associated with the use of human tissues, and was following the appropriate guidance, the director of the committee requested permission from NBAC to translate the *Research Involving Human Biological Materials* report into Japanese.

Chapter Six

RESPONSE TO THE *ETHICAL ISSUES IN HUMAN STEM CELL RESEARCH* REPORT

In November 1998, three reports brought to public attention the scientific and clinical prospects of human stem cell research.[1]

- The first report, by James Thomson and his colleagues at the University of Wisconsin, described the isolation and culture in the laboratory of human embryonic stem cells derived from the blastocyst stage of early embryos donated by couples who had undergone infertility treatments (Thomson et al., 1998).

- The second report, on a study by John Gearhart and his colleagues at The Johns Hopkins University School of Medicine, described the establishment of human embryonic germ cells from primordial gonadal tissue, which was obtained from fetal tissue following elective abortion (Shamblott et al., 1998).

- The third report (Wade, 1998), which appeared in *The New York Times*, described work by scientists at Advanced Cell Technology, Inc. (a company specializing in development of nuclear transfer technologies for human therapeutics and animal cloning) who

---

[1]These three reports were published within days of each other. The first two reports were published in the peer-reviewed scientific literature, while the third report appeared in the lay press and did not undergo the scrutiny of peer review. The first report, by Thomson et al., was published in the November 6, 1998, issue of *Science* and the second report, by Shamblott et al., was published in the November 10, 1998, issue of *Proceedings of the National Academy of Sciences of the United States of America*. These two independent teams of scientists were supported by private funds from the Geron Corporation, a biopharmaceutical company specializing in therapeutic and diagnostic products. The third report (Wade, 1998) appeared in the November 12, 1998, edition of *The New York Times*.

claimed they had fused human somatic cells with enucleated cow eggs to create a hybrid embryo from which they then isolated cells resembling embryonic stem cells.

While stem cells hold the promise of one day being able to grow replacement tissues for people with debilitating or sometimes fatal diseases for which there are currently no good treatments, their use has renewed an important national debate about the ethics of research involving human embryos and fetal tissue and the appropriateness of federal funding for this type of research.

President Clinton responded immediately to these three reports. On November 14, 1998, he wrote to NBAC with two requests: first, that NBAC consider the implications of the reported embryonic stem cells that were part human and part cow and report back to him as soon as possible and, second, that NBAC conduct "a thorough review of the issues associated with . . . human stem cell research, balancing all medical and ethical issues." On November 20, 1998, NBAC responded by letter to the president's first request. The commission concluded that "any attempt to create a child through the fusion of a human cell and a non-human egg would raise profound ethical concerns and should be prohibited." Furthermore, if the fusion of a human cell and a nonhuman egg forms an embryo, "important ethical concerns arise, as is the case with all research involving human embryos." The letter went on to say that if an embryo is not formed, NBAC does not believe that totally new ethical issues arise because "scientists routinely conduct non-controversial and highly beneficial research that involves combining material from human and other species. . . . Combining human cells with non-human eggs might possibly lead some day to methods to overcome transplant rejections without the need to create human embryos."

After nine months of study, NBAC submitted its *Ethical Issues in Human Stem Cell Research* report to the president on September 7, 1999. The report focused primarily on the ethical questions relevant to federal sponsorship of research involving human stem cells. NBAC based its recommendations on the source of human stem cells (adult tissue, cadaveric fetal tissue, embryos remaining after infertility treatments, or embryos created for research purposes by in vitro fertilization or somatic cell nuclear transfer), recognizing that each source raises unique scientific, ethical, and legal issues.

NBAC recommended that federal sponsorship of research involving the derivation and use of human embryonic stem cells and human embryonic germ cells should be limited in two ways: (1) research should be limited to using only embryos remaining after infertility treatments or cadaveric fetal tissue; and (2) sponsorship should be contingent on a system of national oversight and review. Research involving the derivation or use of human embryonic stem cells from embryos created solely for research purposes through either in vitro fertilization or cloning techniques should not be eligible for federal funding. Other recommendations addressed various aspects of human stem cell research, including requirements for informed consent, preventing the buying or selling of embryos or fetal tissue, establishing a National Stem Cell Oversight and Review Panel, requirements for IRB review, and voluntary compliance by the private sector with the same requirements recommended for federally funded researchers.

Since late 1998 when the reports about human embryonic stem cells first appeared, researchers have shown that these cells are capable of becoming virtually all of the specialized cells of the body, and may be able to be used to grow replacement tissues for people with various diseases, including bone marrow for cancer patients, neurons for people with Alzheimer's disease, and pancreatic cells for people with diabetes. This renewable source of human cells with the capability of differentiating into a wide variety of cell types also has broad applications in basic research, such as understanding fundamental events in embryonic development and the causes of birth defects.

Researchers are also investigating ways to isolate stem cells from adult tissue. They have found that some adult stem cells appear to be able to differentiate into tissues other than the ones from which they originated. For example, it has been shown that blood and bone marrow stem cells can differentiate into brain cells, skeletal muscle cells, cardiac muscle cells, and liver cells, and brain stem cells can differentiate into blood cells and skeletal muscle cells. However, current science indicates that embryonic and adult stem cells differ in important ways. For example, researchers have succeeded in turning embryonic stem cells into more than 100 different types of cells, which is far more than what has been accomplished with adult stem cells. Also, embryonic stem cells can reproduce indefinitely in the laboratory, while adult stem cells are difficult or impossible to maintain and expand in culture and eventually die. In addition, re-searchers have been less successful at growing adult stem cells into quantities that would be useful for treatment. Finally, adult stem cells are rare and are often difficult to identify, isolate, and purify. Therefore, there may be several advantages to using embryonic stem cells for the treatment of certain diseases.

The scientific picture became even more complicated in July 2001 when scientists at the Jones Institute for Reproductive Medicine in Norfolk, Virginia, reported that they had created human embryos specifically for the purpose of isolating human embyronic stem cells from them (Lanzendorf et al., 2001), and Advanced Cell Technology reported that its scientists are trying to clone human embryos to ob-tain stem cells (Weiss, 2001b). Scientists at the Jones Institute were the first researchers in the world to report harvesting embryonic stem cells from human embryos that were created specifically for re-search (Weiss, 2001a). Until now, researchers had reported deriving stem cells only from donated embryos that had already been created in fertility clinics but were designated for disposal because they were not needed.

GOVERNMENT RESPONSE

NBAC focused primarily on the ethical questions that are relevant to federal sponsorship of research involving human stem cells because current federal law forbids the use of federal funds for research that creates human embryos or that results in damage or destruction of a human embryo. NBAC recommended that federal funding should be allowed for the derivation and use of human stem cells from cadav-eric fetal tissue and embryos remaining after infertility treatments. NBAC also recommended that DHHS should establish a National Stem Cell Oversight and Review Panel to ensure that all federally funded human stem cell research is conducted ethically.

President Clinton responded to the *Ethical Issues in Human Stem Cell Research* report (as discussed next), and on March 16, 2001, NBAC Chairman Harold Shapiro wrote to President Bush to inform him of the report. Bills were introduced in the 106th and 107th Con-gresses that mentioned NBAC, and several members of Congress mentioned NBAC's report in statements before Congress or in letters to the president. Furthermore, NBAC's work on the ethical issues in human stem cell research was discussed by NBAC commissioners and others several times in congressional testimony. The NIH considered NBAC's recommendations when it developed its guidance on research involving human embryonic stem cells. In addition, a letter from the secretary of HHS to members of Congress stated that NBAC's work would help ensure appropriate oversight of stem cell research. Federal legislation, congressional testimony, guidelines, policy statements, and other relevant documents that refer to NBAC's work on stem cell research are detailed next.

Response from the President of the United States

NBAC met on July 14, 1999, to discuss recommendations of its draft report on stem cell research. These draft recommendations called for federal funding of both the derivation of stem cells from embryos and research on stem cells. The White House immediately released a statement on NBAC's draft recommendations making it clear that the Clinton administration would support only the use of human embryonic stem cells and not their derivation:

> The Clinton Administration recognizes that human stem cell technology's potential medical benefits are compelling and worthy of pursuit, so long as the research is conducted according to the highest ethical standards. NIH is putting in place guidelines and an oversight system that will ensure that the cells are obtained in an ethically sound manner. The President's 1994 ban on the use of Federal funds for the creation of human embryos for research purposes will remain in effect. No other legal actions are necessary at this time, because it appears that human embryonic stem cells will be available from the private sector. Publicly funded research using these cells is permissible under the current Congressional ban on human embryo research.

After receiving NBAC's final *Ethical Issues in Human Stem Cell Research* report on September 7, 1999, President Clinton issued a statement thanking the NBAC commissioners for a "thoughtful" report on "complex and difficult" issues. The president also commended the commissioners for their "thoroughness" in seeking the views and opinions of "virtually every segment of society." The president also stated that "the scientific results that have emerged in just the past few months already strengthen my hope that one day, stem cells will be used to replace cardiac muscle cells for people with heart disease, nerve cells for thousands of Parkinson's patients, or insulin-producing cells for children who suffer from diabetes."

The debate over federal funding for human embryonic stem cell research was renewed during the 2000 presidential campaign when then-candidate George W. Bush expressed his opposition to "federally funded research for experimentation on embryonic stem cells that require live human embryos to be discarded or destroyed." Since then, President Bush also has stated his opposition to research that relies on tissue from induced abortions. Soon after taking office, the president requested that the NIH refrain from funding research on embryonic stem cells until the legal status of the research could be reviewed. President Bush consulted politicians, doctors, religious scholars, and bioethicists on the issue of embryonic stem cell research. In July 2001, the NIH released a report, prepared in response to a request by HHS Secretary Tommy Thompson, which summarized the state of the science on stem cells from embryos, and from fetal and adult tissues. On July 17, 2001, Harold T. Shapiro wrote President Bush to inform him of NBAC's *Ethical Issues in Human Stem Cell Research* report and to offer NBAC's assistance on the question of federal funding for human embryonic stem cell research.

On August 9, 2001, President Bush announced in his first presidential address to the American people that the federal government would fund embryonic stem cell research with existing stem cell lines "where the life-and-death decision has already been made." To be eligible for federal funding, existing stem cell lines must have been derived from excess embryos created solely for reproductive purposes, with the informed consent of the embryo donors, and without any financial inducements to the donors. No federal funds would be used for the derivation of stem cells from newly destroyed embryos or the use of such stem cells, the creation of human embryos for research purposes, or the cloning of human embryos for any purpose. The president also directed NIH to establish a national human embryonic stem cell registry to ensure that federally funded stem cell research meets ethical standards. In addition, the president announced the creation of a new President's Council on Bioethics, to be chaired by University of Chicago professor Leon Kass, "to monitor stem cell research, to recommend appropriate guidelines and regulations, and to consider all of the medical and ethical ramifications of biomedical innovation."

Response from Congress

Legislative Action. Several bills were introduced in the 106th and 107th Congresses to either allow or prohibit federal funding of embryonic stem cell research; however, only one specifically mentioned NBAC. NBAC's recommendations on stem cell research were discussed when a House Resolution was introduced and again when legislation was introduced in the Senate.

On February 2, 2000, Representative Carolyn Maloney (D-N.Y.) and Representative Connie Morella (R-Md.) introduced House Resolution 414, expressing that the House of Representatives "supports Federal funding directed toward human pluripotent[2] stem cell research to further research into Parkinson's disease and other medical conditions." The resolution had 18 cosponsors and was referred to the House Committee on Commerce. Representative Maloney noted NBAC's recommendations on stem cell research when she introduced House Resolution 414, and again in a statement before the House on March 22, 2000. She stated "NBAC points out that Federally funding this [embryonic stem cell] research will allow Federal oversight to ensure this type of research continues ethically."

On January 31, 2000, Senator Arlen Spector (R-Penn.) and Senator Tom Harkin (D-Iowa) introduced S. 2015, the Stem Cell Research Act of 2000. S. 2015 authorizes the secretary of HHS to "conduct, support, or fund research on, or utilizing, human embryos for the purpose of generating embryonic stem cells . . . in accordance with certain restrictions." Human embryonic stem cells may be derived and used in research "only from embryos that otherwise would be discarded [and] that have been donated from in-vitro fertilization clinics with the written informed consent of the progenitors." The bill also specifies three restrictions: (1) the research shall not result in the creation of embryos for research purposes; (2) the research shall not

---

[2]*Pluripotent cells* are cells that are present in the early stages of embryo development that can generate all of the cell types in a fetus and in the adult and that are capable of self-renewal. Pluripotent cells are not capable of developing into an entire organism.

result in the reproductive cloning of a human being; and (3) it shall be unlawful for any person receiving federal funds to knowingly acquire, receive, or otherwise transfer any human gametes (mature male or female reproductive cells) or human embryos for valuable consideration if it affects interstate commerce (i.e., intrastate commerce is controlled at the state level).

While introducing S. 2015, Senator Harkin referred to the NBAC *Ethical Issues in Human Stem Cell Research* report in his remarks. He noted that "NBAC also arrived at the important conclusion that it is ethically acceptable for the federal government to finance research that both derives cell lines from embryos and that uses those cell lines." He went on to say that "NBAC's report presents reasonable guidelines for federal policy." Late in September 2000, Senator Sam Brownback (R-Kan.) blocked a final attempt to get the bill through the Senate. Senators Specter and Harkin reintroduced the bill in the 107th Congress in April 2001 as S. 723, the Stem Cell Research Act of 2001. On June 5, 2001, Representative Jim McDermott (D-Wash.) introduced H.R. 2059 as a companion bill to S. 723.

On April 26, 2001, Senator Brownback and Representative Dave Weldon (R-Fla.) introduced the Human Cloning Prohibition Act of 2001 (S. 790/H.R. 1644), which set civil and criminal penalties for anyone attempting to clone a human being. The bill also contained a "Sense ```of Congress" provision applicable to embryonic stem cell research (see Chapter Three for more details). The Human Cloning Prohibition Act of 2001 directed the president to commission a study, from NBAC or a successor group, on cloning to produce human embryos solely for research. Such embryos could be used to produce stem cells from which transplantable tissue identical to the patient could be derived for therapeutic purposes. However, a revised version of the bill (H.R. 2505), which no longer mentioned NBAC, passed the House on August 1, 2001. However, S. 790 was not passed by the Senate.

Statements by Members of Congress. On June 13, 2001, Senator Orrin Hatch (R-Utah) sent letters to President Bush and Secretary Tommy Thompson expressing his support for federal funding of human embryonic stem cell research. His support of this research is based on it being "conducted in strict accordance with the relevant statutes and the protections set forth in the applicable regulations and guidelines, including those issued by the NIH." In his letter to Secretary Thompson, Senator Hatch notes that the NIH guidelines were based on "an extensive body of earlier work of the National Bioethics Advisory Committee [sic], the Advisory Committee to the Director, NIH, and a special Human Embryo Research Panel. . . ." Senator Hatch urged President Bush to support federal funding of embryonic stem cell research and appealed to the president "to provide the personal leadership required to see to it that [the Bush] Administration will be remembered by future historians as the beginning of the end for such deadly and debilitating diseases as cancer, Alzheimer's, and diabetes."

On July 18, 2001, on the Senate floor, Senator Bill Frist (R-Tenn.) announced his support for federal funding of both embryonic and adult stem cell research:

> After grappling with the issue—scientifically, ethically, and morally—I believe that both embryonic and adult stem cell research should be federally funded within a carefully regulated, fully transparent framework that ensures the highest level of respect for the moral significance of the human embryo.

In his Senate statement, Frist called NBAC a "stakeholder" with a part to play in determining the role of the federal government in stem cell research. He mentioned that NBAC had debated this issue and determined that stem cell research was worthy of federal support. Senator Frist explained that his support was "contingent on the implementation of a comprehensive, strict new set of safeguards and public accountability governing this new, evolving research." He listed ten essential components of a comprehensive framework for stem cell research, which included a ban on federal funding for the derivation of embryonic stem cells, a ban on the creation of embryos for research purposes, a ban on human cloning, a requirement for a rigorous informed consent process, and requirements for a strong public oversight system for research and ongoing, independent scientific and ethical review.

On July 27, 2001, Representative Cliff Stearns (R-Fla.) presented a statement entitled "Against Federal Funding of Embryonic Stem Cell Research" to the House of Representatives. In his statement, Stearns contended that embryonic stem cell research results in the death of the embryo and "violates many of the tenets of the Nuremberg Code and U.S. Law." He argued that an ethical alternative is to use stem cells obtained from adult tissues, and quoted from the *Ethical Issues in Human Stem Cell Research* report to support his argument: "Even the Clinton National Bioethics Advisory Commission said that embryo-destructive research should go forward only 'if no less morally problematic alternatives are available for the research.'" He concluded that federal funds should not be used for embryonic stem cell research.

Congressional Testimony. NBAC's work on the ethical issues in human stem cell research was discussed by NBAC commissioners and was mentioned by others several times in testimony before the Senate Appropriations Subcommittee on Labor, Health and Human Services, Education and Related Agencies:

- While NBAC was still examining the issues associated with human stem cell research, Dr. Eric Meslin, then executive director of NBAC, testified on December 2, 1998, and January 26, 1999, on the status of the commission's report before the Senate subcommittee.

- Dr. Harold Varmus, then director of NIH, also testified at the December 2 and January 26 hearings and mentioned NBAC's work on human stem cell research on both occasions.

- Commissioner Jim Childress testified on November 4, 1999, after the *Ethical Issues in Human Stem Cell Research* report was completed. He described the background and process NBAC used to arrive at its recommendations and summarized some of the report's major recommendations.

- At a hearing on April 26, 2000, Senator Brownback testified that federal funding for embryonic stem cell research is illegal, immoral, and unnecessary. He supported funding for research on adult stem cells instead. Brownback argued that the NIH was inconsistent in allowing the use but not the derivation of embryonic stem cells, and he stated that NBAC made the "more honest and heinous" recommendation that federal funding should be allowed for both derivation and use of embryonic stem cells. Tom Harkin, the ranking member on the Senate subcommittee, argued that limiting stem cell research to the private sector would hinder development of regulations and ethical considerations. He stated that the federal government has already acted responsibly by establishing NBAC and proposing guidelines for stem cell research.

- On September 7, 2000, Gerald Fischbach, M.D., director of NINDS, and Allen Spiegel, M.D., director of the National Institute of Diabetes and Digestive and Kidney Diseases (NIDDK), testified about the newly released NIH guidelines for research using human stem cells. In their testimony, they mentioned that the NIH sought advice from several sources including NBAC, scientists, lawyers, and members of Congress.

NBAC commissioners were also invited to other venues to contribute their insights on ethical issues on stem cell research. On April 10, 2001, Commissioner Patricia Backlar participated in Representative David Wu's (D-Ore.) Stem Cell Research Roundtable in Portland, Oregon, and on September 5, 2001, Jim Childress testified before the Committee on Health, Education, Labor, and Pensions on his personal views on the ethical issues involved in the debate about federal funding of embryonic stem cell research. Even though Commissioner Childress was not testifying on NBAC's behalf, he referred to NBAC's report and recommendations on stem cell research extensively in his testimony.

Response from the Department of Health and Human Services

On February 11, 1999, Representative Jay Dickey (R-Ark.) and 69 other members of Congress wrote to the secretary of HHS, objecting to the legal decision by the general counsel of HHS, which concluded that NIH can fund embryonic stem cell research. Representative Dickey wrote that the general counsel's decision violated both the "letter and spirit of the federal law banning federal support for research in which human embryos are harmed or destroyed." Representative Dickey interpreted the law as not only prohibiting research that destroys an embryo, but also research that depends on the previous destruction of an embryo. Senator Brownback and six other senators wrote a similar letter to HHS Secretary Donna Shalala disagreeing with the HHS position. On February 23, 1999, Shalala responded to Dickey's letter defending the NIH's decision to fund embryonic stem cell research, and explained NIH's plans to develop rigorous guidelines for this research. She also mentioned that NBAC was studying the issue and would provide NIH with advice that will "help ensure appropriate oversight."

Response from the National Institutes of Health

In December 1999, the NIH released its draft guidelines for NIH-sponsored research involving human embryonic stem cells for public comment (National Institutes of Health, 1999). A news release issued on December 1, 1999 ("NIH Publishes . . .," 1999) and a fact sheet accompanying the draft guidelines stated that the NIH "considered advice from the National Bioethics Advisory Commission (NBAC), the public, and the Congress" during the development of its stem cell guidelines. Supplementary information included with the NIH draft guidelines described a public meeting on April 8, 1999, held by the Working Group of the Advisory Committee to the Director of NIH. During the meeting, NBAC Executive Director Eric Meslin discussed the current status of NBAC's deliberations on human stem cell research .

On August 25, 2000, the final *National Institutes of Health Guidelines for Stem Cell Research* were published in the *Federal Register* (National Institutes of Health, 2000). These guidelines "establish procedures to help ensure that NIH-funded research in this area is conducted in an ethical and legal manner." Both an NIH news release, dated August 23, 2000, entitled "NIH Publishes Final Guidelines for Stem Cell Research," and an NIH Fact Sheet on "Human Pluripotent Stem Cell Research Guidelines" (National Institutes of Health, 2001) state that in drafting the guidelines the NIH sought advice from several sources, including NBAC, scientists, patients and patient advocates, ethicists, lawyers, and members of Congress "to help ensure that any research utilizing human pluripotent stem cells is appropriately and carefully conducted."

The NIH guidelines state that while DHHS funds cannot be used for the derivation of stem cells from human embryos, they can be used for research utilizing human embryonic stem cells because those cells are not embryos. This determination of funding eligibility was based on a legal interpretation of federal law that currently restricts the use of DHHS funds for human embryo research. However, a *New York Times* writer characterized NIH's guidelines as "morally inconsistent," citing NBAC's recommendation that both the derivation and use of embryonic stem cells should be eligible for federal funding based on the "close connection in practical and ethical terms between derivation and use of the cells" (Wade, 2000).

NBAC's work on the ethical issues in human stem cell research was mentioned at various meetings of advisory groups to the director of the NIH. A report from the Stem Cell Working Group at the 78th Meeting of the Advisory Committee to the Director of NIH on June 3, 1999, mentioned President Clinton's request to NBAC to review the ethical issues associated with stem cell research (Office of the Director, NIH, 1999a). At the same meeting, Harold Varmus, then director of NIH, emphasized that NIH and DHHS have restricted their analysis to the legal implications of stem cell research, whereas NBAC's analysis has focused on the ethical implications.

At the NIH Director's Council of Public Representatives (COPR) on October 21, 1999, shortly after NBAC released its *Ethical Issues in Human Stem Cell Research* report, Dr. Varmus compared NBAC's conclusion that both the use and derivation of human embryonic stem cells should be eligible for federal funding with the conclusion of the DHHS general counsel that current federal law allows NIH to fund research that uses embryonic stem cells but not the derivation of such cells. An "Update on Stem Cell Research" from the proceedings of the 79th Meeting of the Advisory Committee to the Director of NIH on December 2, 1999, a day after NIH released its draft guidelines for stem cell research, reiterated the DHHS general counsel's conclusion that stem cells are not and cannot become human embryos and therefore stem cell research is eligible for federal funding, adding that NBAC "provided additional support indicating that this position is based on sound ethics" (Office of the Director, NIH, 1999b).

National Institute of Diabetes and Digestive and Kidney Diseases.
The Stem Cell and Developmental Biology Planning Group of the NIDDK is one of a number of ad hoc strategic planning groups that have been established to help NIDDK set priorities for research support in critical scientific and biomedical areas. An NIDDK report entitled *Stem Cell and Developmental Biology Writing Group's Re-*

port—Draft Version 1.0 identified the "sponsorship of activities that promote public education in stem cell research" and the "support of broad-based society discussions of ethical issues arising from stem cell research" as major needs of an integrated program on stem cell biology (National Institute of Diabetes and Digestive and Kidney Diseases, n.d.). The report recognized that stem cell research will "raise ethical concerns among some members of the public," and pointed to NBAC as one of the forums that will discuss the ethical, legal, and social issues in biomedical research and will guide policy decisions in these areas.

RESPONSE FROM PROFESSIONAL SOCIETIES, ORGANIZATIONS, AND FOUNDATIONS

During its deliberations on the ethical issues associated with stem cell research, NBAC wrote to approximately 40 scientific, medical, professional, religious, and health organizations to request their input on the scientific, medical, and ethical issues involved in human stem cell research. NBAC was particularly interested in receiving policy or position statements these organizations had issued relating to the research use of human embryonic or fetal material in general or the research use of embryonic stem cells in particular. Several organizations provided comments for NBAC's consideration during its deliberations on stem cell research.

Although the NBAC report focused primarily on the ethical issues associated with federal funding of research involving human stem cells, NBAC noted that its recommendations might have implications for research on human stem cells conducted by the private sector. In addition, NBAC noted that research with human embryos conducted in the private sector takes place with no federal oversight specific to human embryos as long as federal funds are not involved, and FDA regulations do not apply to such research, and no state law prohibits it. Therefore, NBAC recommended that private sponsors and researchers voluntarily comply with NBAC's recommendations and with the ethical principles underlying its report.

NBAC's deliberations on human stem cell research were followed closely, and its *Ethical Issues in Human Stem Cell Research* report has been cited by several organizations. From the outset, the Biotechnology Industry Organization called on NBAC to address the ethical issues associated with stem cell research. The Endocrine Society supported NBAC's report and recommendations on human stem cell research in its Code of Ethics. The AMA released a report and issued policy supporting NBAC's recommendations. The American Society for Cell Biology, AAMC, and the United States Conference of Catholic Bishops have referred to NBAC's report and recommendations in congressional testimony. AAMC also referred to NBAC's recommendations on human stem cell research in a letter to Congress.

More than 100 university presidents wrote to the secretary of HHS in support of human stem cell research and asserted their agreement with NBAC's recommendations. A report by the AAAS and the Institute for Civil Society (ICS) praised NBAC's role in public policy, whereas publications from the United States Conference of Catholic Bishops were critical of NBAC's recommendations. In addition, many organizations posted information about NBAC's deliberations and recommendations on their Web sites. Congressional testimony, reports, newsletters, statements, and other documents from societies, organizations, and foundations on human stem cell research that mention NBAC are described in more detail next.

Response from the American Association for the Advancement of Science and Institute for Civil Society

In November 1999, AAAS and ICS released a report entitled *Stem Cell Research and Applications: Monitoring the Frontiers of Biomedical Research* (AAAS and ICS, 1999). This report was produced by a working group convened by the two organizations and was informed by discussions that took place during a public meeting hosted by AAAS and ICS on August 25, 1999. At that meeting, Eric Meslin, then NBAC executive director, discussed the key recommendations in the *Ethical Issues in Human Stem Cell Research* report. The AAAS/ICS report recommended that federal funding be allowed for research using embryonic stem cells, but not for the derivation of the cells themselves due to ethical concerns, and the report contended that federal funding would ensure that stem cell research was conducted in an ethical manner and under public scrutiny. The AAAS/ICS report went on to say that NBAC "has demonstrated its legitimate claim to respect for its efforts as a national body to promote public input into social policy related to advances in biomedical research." Other recommendations in the AAAS/ICS report, focused on a need for public education, guidelines for donation of embryos for stem cell research and informed consent, assuring equitable access to the benefits of stem cell research, and intellectual property issues.

Articles in three issues of the *Professional Ethics Report*, a publication of AAAS's Scientific Freedom, Responsibility, and Law Program in collaboration with the Committee on Scientific Freedom and Responsibility and the Professional Society Ethics Group, mentioned NBAC's work in relation to the issue of human stem cell research.

An article in the Spring 1999 issue discussed FDA regulatory controls over human stem cells and mentioned that NBAC was considering the ethical issues associated with the use of human stem cells in federally funded research (Brady, Newberry, and Gerard, 1999).

An article in the Summer 1999 issue announced the release of the NBAC *Ethical Issues in Human Stem Cell Research* report and highlighted several of the recommendations found in the report (Garfinkel, 1999). The article mentioned that NBAC's recommendation for federal funding of both derivation and use of human stem cells was controversial, attracting attention from the White House even before the final report was issued. Specifically, the article points out that the Clinton administration issued a statement making it clear that it would support only the use of human stem cells and not their derivation (as discussed earlier in this chapter).

Lastly, a news article in the Fall 1999 issue compares NBAC's recommendations with the NIH draft guidelines for stem cell research and with recommendations in the AAAS/ICS report *Research and Applications: Monitoring the Frontiers of Biomedical Research* ("NIH Releases . . .," 1999). The article noted that NBAC concluded that federal funds should be used for both the derivation and use of human stem cells, whereas NIH and AAAS/ICS concluded that federal funds should be used only for research with stem cells but not their derivation. The article also pointed out that both NBAC and NIH recommended review of research involving human stem cells at both the local and national levels, whereas AAAS/ICS recommended that no new oversight bodies were necessary for now.

Three articles in *Science & Technology in Congress*, the newsletter of the AAAS Center for Science, Technology, and Congress, discussed NBAC's report and recommendations on stem cell research. The first article, in the July 1999 issue, discussed the Clinton administration's reactions to NBAC's recommendations in a draft version of its report for federal funding of both the derivation and use of human stem cells ("Debate over . . .," 1999). The March 2000 issue highlighted the major difference between the NBAC recommendations and the NIH guidelines, namely that NBAC recommended that federal funding be approved for the derivation of stem cells, whereas the NIH did not ("Battle over . . .," 2000). The article pointed out that this is an important distinction because the NIH decision places responsibility for the derivation of stem cells in the private sector. Finally, a September 2001 article mentioned the *Ethical Issues in Human Stem Cell Research* report in context of the history of stem cell research ("Embryonic Stem Cell . . .," 2001).

NBAC's recommendation on federal funding for research on stem cells derived from embryos left over from fertility treatments was mentioned in *Congressional Action on Research & Development in the FY 2000 Budget*, a report on appropriations for research and development in the federal budget published annually by AAAS (Koizumi and Turner, 2000). Chapter 3 of the report, "Agency R&D Budgets," explained that the Senate appropriations bill for DHHS originally contained a provision that would have allowed federal funding for stem cell research, reflecting NBAC's recommendations on this issue. However, the provision was removed before the bill reached the Senate floor, and the final appropriations bill retained a ban on federal funding for research in which human embryos are destroyed or for the creation of human embryos for research.

Response from the American Council on Education, Association of American Universities, and National Association of State Universities and Land-Grant Colleges

On March 26, 2001, the presidents of the American Council on Education, the Association of American Universities, and the National Association of State Universities and Land-Grant Colleges along with 112 university presidents and chancellors wrote to Tommy Thomson, secretary of HHS, urging him "to permit the current NIH guidelines governing stem cell research to remain in effect" ("ACE, AAU, NASULGC Letter . . .," 2001). The university presidents described the discovery of embryonic stem cells and their potential for curing or treating diseases as being among "the most promising biomedical developments in years."

The letter stated that the university presidents agreed with NBAC's September 1999 report, which concluded that embryonic stem cell research should proceed with public funding. The letter went on to say that the guidelines developed by the NIH will ensure that stem cell research is conducted in an ethical manner. Copies of the letter also were sent to President Bush, Vice President Richard Cheney, and members of Congress. The university presidents' letter also referred to a letter written by a group of 80 Nobel Laureates and sent to President Bush on February 22, 2001, which said that the government should fund embryonic stem cell research. The week before the university presidents sent their letter, 95 members of the House of Representatives, including 5 Republicans, wrote a letter to President Bush urging him to allow federal funding for embryonic stem cell research.

Response from the American Medical Association

The AMA Council on Scientific Affairs issued a report in June 1999 entitled *Embryonic/Pluripotent Stem Cell Research and Funding* that recommended that the AMA support the recommendations found in the NBAC *Ethical Issues in Human Stem Cell Research* report (American Medical Association, 1999c). The AMA report also mentioned NBAC's recommendation that federal funding be allowed for both the derivation and use of stem cells from embryos remaining after infertility treatments, and concluded that the AMA should encourage strong public support of federal funding for stem cell research. At the 1999 AMA Interim Meeting, the AMA House of Delegates, which establishes the main health policies of the association,[3] adopted Policy H-460.917, "Science, Policy Implications, and Current AMA Position Regarding Embryonic/Pluripotent Stem Cell Research

---

[3]AMA House of Delegates' policies are based on professional principles and the scientific viewpoints of many thousands of physicians. House of Delegates policy defines what the AMA stands for as an organization.

and Funding," based on the findings of AMA Council on Scientific Affairs' June 1999 report. The policy states:

> Our AMA: (1) encourages strong public support of federal funding for research involving human pluripotent stem cells (PSC); and (2) supports the recommendations of the National Bioethics Advisory Commission (NBAC) report, Ethical Issues in Human Stem Cell Research, September 1999.

An article in the August 2, 1999, *American Medical News* reviewed NBAC's recommendations on human stem cell research (Gianelli, 1999). The article gave details of NBAC's final deliberations on stem cell research, relating that NBAC adopted an "intermediate" position that an embryo merits respect as human life, but not at the level generally accorded to persons. The article highlighted several of NBAC's key recommendations, including that research involving stem cells from embryos left over from infertility treatments and research involving germ cells from aborted fetuses should be federally funded; research using embryos created solely for research purposes, through either in vitro fertilization or cloning techniques, should not be federally funded; a national review panel and public registry should be established; and proper informed consent procedures must be followed for embryo and tissue donors.

Response from the American Society for Cell Biology

The American Society for Cell Biology cited the *Ethical Issues in Human Stem Cell Research* report in congressional testimony. On April 26, 2000, Lawrence Goldstein represented the society at a hearing of the Senate Labor, Health and Human Services, and Education Subcommittee. In his comments in support of S. 2015, the Stem Cell Research Act of 2000, Dr. Goldstein described NBAC's process of deliberation and the commission's recommendations on human stem cell research. He also stated that the American Society for Cell Biology agrees with NBAC that "relying on cell lines that might be derived exclusively by a subset of privately funded researchers who are interested in this area could severely limit scientific and clinical progress."

Response from the Association of American Medical Colleges

AAMC followed NBAC's deliberations on stem cell research very closely. For example, AAMC summarized each of the NBAC meetings on stem cell research and the congressional testimony given by NBAC commissioners and staff in *AAMC Washington Highlights*, the association's weekly news bulletin on legislative and regulatory events in Washington, D.C. Several issues of *AAMC Issue Briefs* also summarized NBAC's role in the debate on this subject.

Representatives of AAMC also referenced the NBAC report on stem cell research in congressional testimony. A February 19, 1999, letter to Congress, written by AAMC in collaboration with 68 other scientific, medical, academic, patient advocacy, and industry organizations, supported DHHS's determination that current law permits the use of federal funds for research with embryonic stem cells, and assured Congress that NBAC's advice would be considered in the drafting of the NIH guidelines for stem cell research.[4] Lastly, in a comment letter on the NIH proposed stem cell guidelines ("AAMC Comment Letter . . .," 2000), the AAMC suggested that if the current federal funding restrictions on embryo research were ever lifted, the NIH should revise its guidelines to allow federal funding for the derivation of stem cells from embryos as recommended by NBAC.

Response from the Biotechnology Industry Organization

On November 12, 1998, just after the three reports about human embryonic stem cells mentioned at the beginning of this chapter were published, BIO issued a news release urging NBAC to "review all ethical issues raised by the recently announced research and after reasoned discussion, to make recommendations to the president on how best to address the implications" ("BIO Urges . . .," 1998). BIO recognized that while embryonic stem cell research holds enormous potential for both basic research and health, it also raises ethical questions that "require thorough and open national discussion." BIO was also concerned that lawmakers may introduce state and federal legislation that might inadvertently impede biomedical research.

---

[4]A copy of the letter sent to Congress by AAMC and the other organizations can be found in American Medical Association (1999c, Appendix I).

Response from the Center for Bioethics and Human Dignity

The Center for Bioethics and Human Dignity, founded in 1993 by leading Christian bioethicists, released a statement on July 1, 1999, entitled "On Human Embryos and Medical Research: An Appeal for Legally and Ethically Responsible Science and Public Policy" (Center for Bioethics and Human Dignity, 1999). The statement addresses the legal, ethical, and scientific issues associated with human embryo stem cell research, and concludes that "human stem cell research requiring the destruction of human embryos is objectionable on legal, ethical, and scientific grounds." The statement also declares that "destruction of human embryonic life is unnecessary for medical progress, as alternative methods of obtaining human stem cells and of repairing and regenerating human tissue exist and continue to be developed." The statement refers to NBAC's May 6, 1999, draft version of its stem cell research report, noting that while NIH determined that the use of embryonic stem cells could be separated from the derivation of embryonic stem cells, NBAC concluded that researchers using stem cells derived from embryos may be implicated in the destruction of embryos.

Response from the Endocrine Society

The Code of Ethics of the Endocrine Society, developed by the society's Ethics Advisory Committee and approved in January 2001, was designed to include "general principles and guidelines to assist with emerging ethical issues" (Endocrine Society, 2001). The Code of Ethics addresses the responsibilities of the society and its members in several areas, including conflict of interest, informed consent, genetic research, clinical trials, and human stem cell research. The Endocrine Society Code of Ethics supports the report and recommendations of NBAC on human stem cell research, as well as the NIH guidelines on stem cell research and the American Society for Reproductive Medicine guidelines on the use of gametes and embryos for research. The Endocrine Society encourages the use of human stem cells for both basic research and clinical applications.

Response from the Joint Steering Committee for Public Policy

The Joint Steering Committee (JSC) for Public Policy is a coalition of three basic biomedical research societies—The American Society for Cell Biology, the American Society for Biochemistry and Molecular Biology, and the Genetics Society of America. JSC represents more than 20,000 researchers in the fields of genetics, cell biology, biochemistry, and molecular biology. JSC was formed in 1990 to bring scientists together to advocate for federal funding for basic biomedical research. The JSC Web site (http://www.jscpp.org/Stemcell.htm) has tracked the stem cell issue and contains information about NBAC, including its report on stem cell research, the deliberations that led up to the report being issued, and congressional testimony given by NBAC commissioner Jim Childress.

Response from the National Academies

Two boards at the National Academies—the National Research Council's Board on Life Sciences and the Institute of Medicine's Board on Neuroscience and Behavioral Health—formed the Committee on the Biological and Biomedical Applications of Stem Cell Research to address the biology of human stem cells and their therapeutic potential. The committee's report, *Stem Cells and the Future of Regenerative Medicine,* was based on the proceedings of a June 22, 2001, workshop, during which scientists addressed several questions on the biology of human stem cells and their potential therapeutic uses, and philosophers, ethicists, and legal experts presented ethical and other arguments relevant to public policy considerations on stem cell research (National Academies, 2001).

The committee's report covered some of the discussion on religious issues in the NBAC report on human stem cell research. The committee's report also pointed out that every federal commission, including NBAC, which has addressed research on human embryos and fetuses has "called for respect for these entities as forms of human life." The committee's report concluded by making several recommendations about stem cell research, including recommendations for research with both embryonic and adult human stem cells, the development of new stem cell lines in the future, public funding of stem cell research to ensure that the therapeutic benefits of stem cells are realized, and a national advisory group to oversee research on human embryonic stem cells.

Response from the Society for Developmental Biology

The Society for Developmental Biology has published two documents focusing on embryonic stem cells that mention the NBAC report. "Update on Human Pluripotent Embryonic Stem Cell Research" (Chow, 1999) mentions the president's request for advice from NBAC on stem cell research and mentions the first two meetings that NBAC had on the issue. "Human Embryonic Stem Cell Research—Where Does It Stand Now?" (Society for Developmental Biology, 1999) comments on the public nature of NBAC's deliberations. It recounts that on several occasions the media reported passages from drafts of the NBAC reports and candid comments made by NBAC commissioners as "decisions" of the commission "creating unwarranted confusion, [and] even leading to the release of official White House statements."

Response from the United States Conference of Catholic Bishops

The United States Conference of Catholic Bishops refers to the *Ethical Issues in Human Stem Cell Research* report several times on its Web site (http://www.usccb.org), and has cited the report in its newsletters and publications, in Catholic newspapers, in statements, and in congressional testimony. The Conference of Catholic Bishops is critical of NBAC's recommendation to federally fund the derivation of stem cells from human embryos and to federally fund their use in research. The group is also critical of the terminology used in the NBAC report, arguing that the term "derivation" is "an innocuous way to describe killing a human embryo."

The Conference of Catholic Bishops has also criticized the NIH for arguing that the derivation of stem cells from human embryos can be separated from their use in research, and therefore, it can fund research using stem cells derived from human embryos if the stem cells were made by privately funded researchers. The U.S. Conference of Catholic Bishops agrees with NBAC's conclusions that "human embryos deserve respect as a form of human life," and that there is no ethical distinction between the derivation and the use of embryonic stem cells. However, this line of reasoning also led NBAC to conclude that both the derivation and the use of embryonic stem cells should be eligible for federal funding. But the Conference of Catholic Bishops argued that because a human embryo is a human life, it is not permissible to kill it to obtain stem cells. The bishops also criticized NIH's claim that federally funded researchers can use embryonic stem cells derived by privately funded researchers as being "disingenuous" and argued that NIH had "only succeeded in producing evasive legal interpretations, questionable science, and incoherent ethics."

INTERNATIONAL RESPONSE

NBAC's report and recommendations on human stem cell research have been cited often in discussions by groups in other countries and by international organizations, and they have informed the public policy debate on this issue worldwide. NBAC's recommendations were discussed in reports on stem cell research by an Expert Advisory Group to the government of the United Kingdom, and by the Nuffield Council on Bioethics. Both the Expert Advisory Group and the Nuffield Council recommended that stem cell research with embryos created through somatic cell nuclear transfer (i.e., "therapeutic cloning") should be allowed. In contrast, NBAC recommended that therapeutic cloning should not be eligible for government funding.

The UNESCO International Bioethics Committee, the Canadian Institutes of Health Research, the Bioethics Committee of the Council for Science and Technology in Japan, and the European Group on Ethics in Science and New Technologies to the European Commission have all cited the *Ethical Issues in Human Stem Cell Research* report in various publications. References to NBAC's report and recommendations on stem cell research by other countries and international groups are detailed next.

Response from the United Kingdom

In June 1999, in response to a December 1998 report entitled *Cloning Issues in Reproductive Science and Medicine* by the UK's HFEA and HGAC, the Government of the United Kingdom asked its Chief Medical Officer (CMO) to establish an Expert Advisory Group to assess developments in stem cell research and nuclear transfer technology and to give advice on whether new areas of research with human embryos should be permitted.

In June 2000, the CMO's Expert Group Reviewing the Potential of Developments in Stem Cell Research and Cell Nuclear Replacement to Benefit Human Health issued its *Stem Cell Research: Medical Progress with Responsibility* report (Chief Medical Officer's Expert Group . . ., 2000). The group's report mentioned NBAC's findings on cloning human beings as well as those on stem cell research, which included: (1) the creation of a human being through somatic cell nuclear transfer should be prohibited; (2) federal funding should be allowed for the derivation and use of stem cells from aborted fetuses and embryos remaining after infertility treatment; (3) the creation of embryos through either in vitro fertilization or somatic cell nuclear transfer should not be eligible for federal funding; and 4) a National Embryonic Stem Cell Oversight and Review Panel should be established.

In comparison with NBAC's recommendations, the Expert Advisory Group recommended that research using embryos created by either in vitro fertilization or somatic cell nuclear transfer should be permitted and should be subject to the controls of the Human Fertilisation and Embryology Act of 1990. The Expert Advisory Group also recommended that the mixing of human somatic cells with the eggs of any other animal species should not be permitted, and that the transfer of an embryo created by somatic cell nuclear transfer into the uterus of a woman should remain a criminal offense.

In August 2000, the Government of the United Kingdom accepted the Expert Advisory Group's recommendations in full and introduced for debate and a free vote the legislation necessary to extend the purposes for which embryos can be used in research. The House of Commons on December 18, 2000, and the House of Lords on January 22, 2001, voted to extend the Human Fertilisation and Embryology Act to allow embryonic stem cell research, including stem cell research with embryos created through somatic cell nuclear transfer. However, the UK government decided that the HFEA could not grant licenses for human embryonic stem cell research until a House of

Lords Select Committee had reported on the issues associated with human cloning and stem cell research.[5]

The Nuffield Council on Bioethics issued a discussion paper entitled "Stem Cell Therapy: The Ethical Issues" on April 6, 2000 (Nuffield Council on Bioethics, 2000b). The Nuffield Council recommended that the regulation in the United Kingdom on embryo research be amended to permit research on embryonic stem cells to develop new therapies. The Nuffield Council concluded that donated embryos from in vitro fertilization treatments could be used for embryonic stem cell research, but there were no compelling reasons to allow additional embryos to be created merely to increase the number of embryos available for embryonic stem cell research or therapy. However, on the question of using cloning technology to create embryonic stem cells, the Nuffield Council concluded that "the proposed creation of embryos using SCNT [somatic cell nuclear transfer] for research into the derivation of stem cells offers such significant potential medical benefits that research for such purposes should be licensed." In its discussion of informed consent, the Nuffield Council endorsed NBAC's recommendations pertaining to consent for the donation of embryos for stem cell research and restrictions on designating recipients of those donations.

The Public Health Genetics Unit (PHGU), which is funded by the UK's National Health Service, was established in 1997 "to keep abreast of developments in molecular and clinical genetics, and their ethical, legal, social and public health implications." On the PHGU Web site, an on-line publication providing information on "Stem Cells and Cloning" states that NBAC "extensively reviewed" the issue of stem cell research in its report, and recommended that an exception should be made to the ban on federal funding of embryo research to allow the derivation and use of human embryonic stem cells from embryos remaining after infertility treatments (Public Health Genetics Unit, n.d.). The PHGU concluded, "In the current political climate in the U.S., it seems unlikely that the NBAC's recommendations will be adopted."

---

[5]The House of Lords did report on these issues. The report is available at http://www.publications.parliament.uk/pa/ld/ldstem.htm (as of April 2003).

Response from the European Parliament

In July 2000, the European Parliament Directorate General for Research issued a final study entitled *The Ethical Implications of Research Involving Human Embryos* (European Parliament . . ., 2000). This study examined the possible policy options for human embryo research in Europe. It analyzed the existing legal positions among European member states, provided a comparative analysis of policies adopted elsewhere, and explored the ethical arguments relating to the moral status of the embryo. NBAC's use of the terminology "zygote," "embryo," and "fetus" was discussed in the context of the definition of an "embryo." The European Parliament study also examined recent public policy debates on the issue of human stem cell research and cloning, and made several mentions of NBAC's report and recommendations on stem cell research. The study points out that the focus of the NBAC report—the issue of federal funding for embryonic stem cell research—is of particular relevance to the European Parliament because in 1998 the Parliament attempted to attach an amendment to the Framework V Programme[6] legislation to prohibit funding of human embryo research.

Response from the European Group on Ethics in Science and New Technologies to the European Commission

The European Group on Ethics in Science and New Technologies (EGE) to the European Commission issued "Opinion No. 15, Ethical Aspects of Human Stem Cell Research and Use" on November 14, 2000 (European Group on Ethics . . ., 2000). The EGE cited NBAC's *Ethical Issues in Human Stem Cell Research* report in that opinion document. The EGE based its opinion on the ethical issues raised by human stem cell research and use depending on whether the stem cells were derived from adult tissue, umbilical cord blood, fetal tissue, or embryos.

Because derivation of stem cells from embryos raises the issue of the moral status of the human embryo, the EGE left the decision to each

---

[6]Fifth Framework Programme of the European Community for research, technological development and demonstration activities (1998–2002); see http://europa.eu.int/comm/research/fp5.html.

member state of the European Union to forbid or to authorize research with embryos remaining after infertility treatments. The EGE found the creation of embryos for research purposes by in vitro fertilization to be ethically unacceptable when excess embryos are available, and it found the creation of embryos by somatic cell nuclear transfer to be scientifically premature. Section 1.15 of Opinion No. 15, which deals with the U.S.'s approach to embryo and stem cell research, points out that, in contrast to Europe, there is a sharp distinction between the public and private sectors in the United States (e.g., there is a congressional ban on the use of federal funds for embryo research but no such legislation governing the private sector).

Response from the International Bioethics Committee of the United Nations Educational, Scientific and Cultural Organization

At a meeting at UNESCO headquarters in Paris in April 2000, a working group of the IBC addressed the issue of whether it is ethically acceptable to derive stem cells from human embryos for therapeutic research. The IBC drafted a report based on the discussions that this meeting generated. The draft report was discussed at the Seventh Session of the IBC in Quito, Ecuador, on November 7–9, 2000, and a final document was prepared following the meeting of the Extended Working Group on the Ethical Aspects of Embryonic Stem Cell Research from January 29 through February 2, 2001, at UNESCO Headquarters. The final report, *The Use of Embryonic Stem Cells in Therapeutic Research: Report of the IBC on the Ethical Aspects of Human Embryonic Stem Cell Research* (International Bioethics Committee, 2001), was issued on April 6, 2001.

The IBC report includes sections on the science and applications of embryonic stem cells, philosophical and religious views on the issue of stem cell research, ethical arguments for and against and restraints on such research, and the various sources of embryonic stem cells. In its discussion of the existing international and national laws and regulations governing human embryo research, the report notes that in the United States federal funding of embryo research is prohibited, but regulation of embryo research in the private sector is left up to the discretion of each state.

The IBC report cites the NBAC *Ethical Issues in Human Stem Cell Research* report and specifically mentions the recommendation that federal funding should be allowed for research using stem cells derived from embryos remaining after infertility treatments but not embryos created for research purposes by either in vitro fertilization or somatic cell nuclear transfer. The IBC report states that human embryonic stem cell research needs to be debated at the national level for each country to determine whether or not to allow such research. The report concludes: "If human embryonic stem cell research is allowed, steps should be taken to ensure that such research is carried out within the framework of a State-sponsored regulatory system that would give due weight to ethical considerations, and set up appropriate guidelines."

Response from Canada

On March 29, 2001, the Canadian Institutes of Health Research (CIHR) issued "A Discussion Paper: Human Stem Cell Research: Opportunities for Health and Ethical Perspectives," which was developed by an ad hoc working group of the CIHR (Canadian Institutes of Health Research, 2001). The discussion paper provides a regulatory framework for funding of human embryonic stem cell research by the institutes The working group recommended that: (1) research on existing human stem cell lines should be eligible for funding by CIHR; (2) research to generate new stem cells derived from human fetal tissue or from human embryos that remain after infertility treatment should be eligible for funding; (3) creating human embryos for the purpose of generating stem cell lines should not be permitted; and (4) CIHR should place a moratorium on the use of somatic cell nuclear transfer to create stem cells, research in which human embryonic stem cells are used to create or contribute to a human embryo, and research that involves combining human embryonic stem cells with animal embryos or vice versa.

The preamble to the CIHR paper explains that the working group did not attempt an in-depth analysis of ethical issues such as had been done in the United States by NBAC and NIH, in the United Kingdom by the Nuffield Council on Bioethics and the Department of Health, by the Netherlands Ministry of Health, Welfare and Sport, and by the European Commission. A section of the paper titled "The World-

Wide Regulatory Situation for Stem Cell Research" provides an overview of the work done by "expert working groups" to "establish the ethical framework for stem cell research." NBAC's recommendations on stem cell research are discussed as part of the overview. The CIHR paper explains that NBAC felt that the U.S. congressional ban on federal funding of embryo research "conflicted with the ethical goals of medicine involving healing, prevention, and research, and that it was important that federally funded researchers not be scientifically limited by having to rely on ES [embryonic stem] cells derived with private funds."

NBAC is also mentioned in a prior CIHR document, "Working Paper: The Ethics Mandate of the Canadian Institutes of Health Research: Implementing a Transformative Vision" (Canadian Institutes of Health Research, 1999). This document responded to a request of the Interim Governing Council of CIHR for advice on the role that ethics plays across the broad spectrum of CIHR activities. A section on ethics in the working paper raises the question of what ethical norms should govern research or policy on cloning and stem cells. In that section, NBAC's reports on cloning of human beings and stem cell research are referenced.

The Canadian Biotechnology Advisory Committee (CBAC) advises the Biotechnology Ministerial Coordinating Committee on the broad policy issues associated with the ethical, social, regulatory, economic, scientific, environmental, and health aspects of biotechnology. The CBAC *Annual Report 1999–2000* references the NBAC *Ethical Issues in Human Stem Cell Research* report in the context of NIH's guidelines being based on NBAC's recommendation that embryonic stem cell research should be allowed to be conducted using federal funds (Canadian Biotechnology Advisory Committee, 2001).

Response from Germany

Germany's Institut für Wissenschaft und Ethik published the Executive Summary of *Ethical Issues in Stem Cell Research* in its 2000 edition of *Jahrbuch für Wissenschaft und Ethik* (Yearbook for Science and Ethics). The Institut für Wissenschaft und Ethik "found it very important to address the attention of the German public and scientific community to bioethical research in the USA," and indicated that it would consider publication of any further NBAC reports in future editions of its journal (Hübner, 2000).

Response from Japan

In March 2000, the Human Embryo Research Subcommittee of the Bioethics Committee of the Council for Science and Technology (CST),[7] an advisory body to the Japanese prime minister, released a report on embryonic stem cell research entitled *Basic Viewpoints on Human Embryo Research Which Centers on Research Involving Human Embryonic Stem Cells* (Council for Science and Technology, 2000). The Human Embryo Research Subcommittee recommended that research with human embryonic stem cells should be allowed under strict conditions, and on March 13, 2000, the Bioethics Committee of the CST decided to approve such research under specific conditions: (1) researchers would only be allowed to use embryos remaining after infertility treatments and only with the informed consent of the donors; (2) the sale or purchase of fertilized eggs would be prohibited; and (3) the creation of human embryos by in vitro fertilization or by somatic cell nuclear transfer and the creation of human embryonic germ cells from fetal tissues would be prohibited. An unofficial English translation of a draft of the Human Embryo Research Subcommittee report in Japanese, dated February 2, 2000, cites the NBAC *Ethical Issues in Human Stem Cell Research* report and its recommendation that federal funds should be allowed for the derivation and use of human embryonic stem cells created from embryos remaining after infertility treatments.

Response from Slovakia

An article in the *Journal for Medical Ethics and Bioethics* (*Medicínska Etika & Bioetika*) about NIH's draft guidelines on stem cell research concludes that the guidelines "assume a position on the status of the human embryo that is morally inadequate," and that NIH should not fund human embryonic stem cell research (Sotis, 2000). The article also raises the concern that by sanctioning research that is depen-

---

[7]CST was superseded by the Council for Science and Technology Policy after a reorganization of government ministries and agencies on January 6, 2001.

dent on the destruction of human embryos, NIH is morally complicit in the destruction of the embryos. The article references the *Ethical Issues in Human Stem Cell Research* report, saying that NBAC's position that the human embryo and fetus "deserve respect as forms of human life" is problematic. However, the author of the article stated that a formal review of the NBAC report was not within the scope of the article.

RESPONSE TO THE *ETHICAL AND POLICY ISSUES IN INTERNATIONAL RESEARCH: CLINICAL TRIALS IN DEVELOPING COUNTRIES* REPORT

Over the past ten years or more, there has been a significant increase in the amount of clinical research that the United States conducts or sponsors in other countries, particularly in developing countries. As the pace and scope of international research have increased, adequate protection of the rights and welfare of individuals who participate in clinical trials has emerged as a critical issue in international research ethics.

This issue came to light in 1997 when controversy arose over a series of clinical trials conducted in Africa, Asia, and the Caribbean that were aimed at finding a less-complicated, less-expensive, and shorter treatment regimen to lower the rate of maternal-to-infant transmission of HIV in developing countries (Angell, 1997; Lurie and Wolfe, 1997). The concern with these trials was that participants in the control group were given a placebo instead of the existing effective treatment (continuously administered AZT, generic name zidovudine or ZDV, a class of anti-HIV drugs) and were being treated differently than control groups in developed countries where it would be considered unethical to withhold an effective treatment.

In addition, several international efforts were underway in recent years to revise and develop guidelines on the ethical conduct of international research. These efforts included revisions of the World Medical Association's *Declaration of Helsinki*, the WHO's *Operational Guidelines for Ethics Committees That Review Biomedical Research*, and the Council for International Organizations of Medical Sciences (CIOMS) *International Ethical Guidelines for Biomedical Research Involving Human Subjects*.

These issues prompted NBAC to address the ethical, legal, and policy issues that arise when research subject to U.S. research regulations is sponsored or conducted in other countries. NBAC looked at the recruitment of research participants, informed consent, and the risks and potential benefits of conducting research. NBAC also analyzed many national and international guidelines and statements to enhance the protections of research participants in international research. Finally, NBAC focused on the obligations of private and public sponsors to participants, communities, and countries throughout the research process.

On September 29, 2000, NBAC released a draft report on this subject for public comment and received 132 responses, 60 of which were from respondents outside the United States. On April 18, 2001, after 18 months of study, NBAC submitted its report *Ethical and Policy Issues in International Research: Clinical Trials in Developing Countries* to President Bush. The report focused on the ethical issues that arise when clinical trials that are subject to U.S. regulation are sponsored or conducted in developing countries.

The report made 28 recommendations that called for a range of changes in the conduct of research in developing countries to ensure that participants are adequately protected and to add safeguards to the way that U.S. researchers and sponsors conduct clinical trials in developing countries. Specifically, NBAC recommended that clinical trials conducted by U.S. interests abroad should be directly relevant to the health needs of the population of the host country. Other recommendations addressed the choice of research designs, community involvement, the informed consent process, ethics review, and post-trial access to successful research products.

GOVERNMENT RESPONSE

In *Ethical and Policy Issues in International Research: Clinical Trials in Developing Countries*, NBAC made several recommendations directed to the U.S. government. Specifically, NBAC recommended that the U.S. government should not sponsor or conduct clinical trials, either domestic or abroad, that do not provide the ethical protections for individuals participating in the research outlined in its report. NBAC also recommended that the FDA should not accept data obtained from clinical trials that do not provide the substantive ethical protections outlined in the report. In addition, NBAC recommended changes in U.S. research regulations regarding the informed consent process, and recommended that NIH, CDC, and other U.S. agencies should support research on the informed consent process.

NBAC's work on the ethical conduct of U.S.-sponsored research in developing countries was referenced by two NIH institutes that sponsor and/or conduct clinical trials in developing countries. Details about the response of these two institutes to NBAC's *Ethical and Policy Issues in International Research: Clinical Trials in Developing Countries* report follow.

Response from the National Institutes of Health

John E. Fogarty International Center for Advanced Study in the Health Sciences. The Forty-Fifth Meeting of the Advisory Board of the John E. Fogarty International Center for Advanced Study in the Health Sciences (FIC; http://www.fic.nih.gov/) on May 16, 2000, addressed the "Bioethics of Clinical Research Internationally" (John E. Fogarty International Center, 2000). During the discussion, Dr. Gerald T. Keusch, director of FIC, told the advisory board that FIC had expressed its concern to NBAC about the need for broad international input on the issues NBAC was addressing in its report on international research. He also told the board that FIC had been attending NBAC meetings and had been asked to provide comments on the issues to be addressed in the NBAC report.

National Institute of Allergy and Infectious Diseases. The NIAID Division of AIDS Web page on "Ethical Issues in HIV Prevention Trials" mentions that NBAC has developed guidelines for the ethical conduct of U.S.-sponsored research in developing countries (The NIAID Division of AIDS, n.d.). The Web page states that NBAC's report, along with several other sources of guidance, including the Declaration of Helsinki, the Belmont Report, guidance from CIOMS, WHO, and the Joint United Nations Programme on HIV/AIDS (UNAIDS), and U.S. federal regulations, "provide the foundation upon which DIADS-supported prevention trials are designed and implemented." The Web page also states that these sources of guidance provide the basis for the choice of "control" interventions, methods to ensure informed consent, benefits to research participants, and post-trial access to successful research products.

RESPONSE FROM PROFESSIONAL SOCIETIES, ORGANIZATIONS, AND FOUNDATIONS

Several of the recommendations in the NBAC *Ethical and Policy Issues in International Research* report are directed to all sponsors and researchers of clinical trials in developing countries who are subject to U.S. regulations. This includes not only research sponsored or conducted by the federal government, but all research sponsored or conducted by the private sector that is subject to regulation by the FDA or other relevant federal agency. Therefore, the recommendations in the NBAC report are relevant to many of the U.S. sponsors and researchers who perform clinical trials in developing countries.

A report by the AMA Council on Ethical and Judicial Affairs contains a section on NBAC's report and recommendations on international research. Public Citizen, a national nonprofit "watchdog" group (http://www.citizen.org/), followed NBAC's deliberations quite closely, providing commentary on the contents of the report and writing letters to Harold Varmus, then director of NIH, and President Clinton. In addition, NBAC's report and recommendations on international research were mentioned in news articles in publications of the AMA and AAAS. Details about the responses of these organizations to the NBAC report *Ethical and Policy Issues in International Research* are provided next.

Response from the American Association for the Advancement of Science

Two articles in the *Professional Ethics Report*, a publication of the AAAS Scientific Freedom, Responsibility, and Law Program, discussed NBAC's report and recommendations on international research. The first of the two articles ("Human Trials Protected . . .," 2000) featured NBAC's draft report that had just been released. The article highlighted several of NBAC's draft recommendations, and concluded by stating, "Although its guidelines may be influential, NBAC does not have the authority to monitor trials." The second article ("NBAC Issues Report . . .," 2001) featured the release of NBAC's final *Ethical and Policy Issues in International Research* report. This article also highlighted several of NBAC's recommendations, and mentioned Public Citizen's criticisms of the report (Public Citizen's response to NBAC's report appears below).

Response from the American Medical Association

An article in the AMA's *American Medical News* on February 12, 2001, discussed NBAC's report on international research, commenting that NBAC's recommendations "have the potential to significantly influence dialogue surrounding this issue here in the United States" ("Standards for International . . .," 2001). The article both praised and criticized NBAC's recommendation on standards for treating control group participants. The article also pointed out that both the NBAC report and the World Medical Association's Declaration of Helsinki (see http://www.wma.net/e/policy/17-c_e.html) contain language ensuring that participants in developing countries are not only protected from harm, but actually benefit in some way from participating in research.

In a June 2001 report entitled *Ethical Considerations in International Research*, the American Medical Association CEJA analyzed the ethical dilemmas U.S. physicians face in their participation in research conducted in other countries (American Medical Association, 2001). The CEJA report discussed the *Ethical and Policy Issues in International Research: Clinical Trials in Developing Countries* report, stating that "the extensive analysis of the Commission provides a valuable contribution to understanding the ethical issues at stake from a U.S. perspective."

NBAC's division of ethical requirements for the protection of human research participants into substantive and procedural requirements was highlighted by the CEJA report, along with several of NBAC's recommendations. In addition, the recommendations in the CEJA report were very similar to the recommendations made by NBAC, including choosing a scientifically sound research design; minimizing risk and maximizing benefits for research participants; involving representatives of the host country; obtaining IRB approval; ensuring voluntary informed consent; ensuring that the research is responsive to the health needs of the host country; and encouraging research sponsors to continue to provide beneficial study interventions to all study participants at the conclusion of the study.

Response from Public Citizen

Public Citizen's interest in NBAC's work on international research stems from its involvement in publicizing the controversy over clinical trials involving maternal-to-infant transmission of HIV in developing countries. Public Citizen's coverage of NBAC's work and its response to NBAC's report and recommendations were extensive.

In April 1997, Public Citizen wrote to Donna Shalala, then secretary of HHS, strongly urging that nine clinical trials, funded by NIH or CDC, involving HIV-positive pregnant women in developing countries be changed so that all women in the study received an active treatment instead of a placebo or other unproven treatment (Public Citizen, 1997b). Public Citizen called it "highly unethical" to expose large numbers of women to placebo or treatments not proven effective and to needlessly leave infants at risk of contracting HIV. In June 1997, Public Citizen wrote to President Clinton urging him to take action by ordering changes in the clinical trials, or at least to refer this "clearly urgent and controversial" matter to NBAC for review (Public Citizen, 1997a).

In November 1999, Public Citizen wrote to Harold Varmus, then director of NIH, seeking clarification on NIH's "position on future studies of HIV-infected pregnant women." This correspondence was prompted by Dr. Jack Killen's testimony at the September 16, 1999, NBAC meeting about the acceptability of observational studies of HIV-positive pregnant women in which the HIV transmission rate to their offspring was determined without providing these women with treatments known to be effective. Public Citizen urged Dr. Varmus "to immediately and publicly dissociate [himself] from Dr. Killen's comments and clearly state that NIH is now opposed to all observational studies of perinatal HIV transmission and will not fund any such studies."

On February 22, 2001, Public Citizen wrote a letter to Tommy Thompson, secretary of HHS, to object to the FDA's consideration of a placebo-controlled trial of a surfactant for premature infants with Respiratory Distress Syndrome in Latin America, calling the proposed trial "unethical" and "exploitive" (Public Citizen, 2001a). The letter referenced NBAC's draft report on international research ethics, calling it "grossly deficient" for allowing an exception that would allow control groups in clinical trials to be given care that is less than would be available under ideal circumstances. On April 30, 2001, upon release of NBAC's final report to the public, Public Citizen released a statement entitled "National Bioethics Advisory Commission Report Dangerously Weakens International Protections: Report Provides Less Protection for Participants in International Research Than Declaration of Helsinki" (Public Citizen, 2001b). The statement identified three critical areas in which NBAC's recommendations were "inadequate" and weaker than those required by the Declaration of Helsinki: (1) treatment for patients during clinical trials; (2) availability of treatment at trial completion; and (3) American ethical review of American research.[1] The statement claimed that NBAC's report would "put the U.S. in the unenviable position of endorsing ethics standards for American researchers that are lower than those required by the World Medical Association's Declaration of Helsinki."

Not all of Public Citizen's comments about the NBAC report *Ethical and Policy Issues in International Research* were negative. In a July 13, 2001, letter to Dr. Jack Bryant, president of CIOMS, Public Citizen referenced NBAC's report several times to support its criticism of a draft of the CIOMS International Ethical Guidelines for Biomedical Research Involving Human Subjects. For example, Public Citizen pointed to NBAC's conclusion that it is not permissible to use a placebo control for the study of a life-threatening disease for which an established, effective treatment exists.

INTERNATIONAL RESPONSE

The NBAC report *Ethical and Policy Issues in International Research: Clinical Trials in Developing Countries* did not specifically make any recommendations directed to other countries or international organizations. However, it did make recommendations to U.S. sponsors and researchers about enhancing research collaborations between developing and developed countries by (1) building host country ca-

---

[1]Public Citizen had previously stated its disappointment with NBAC's recommendations on the requirement for ethical review by an ethics review board in the host country and by a U.S. IRB. The statement was made in a letter to Harold T. Shapiro, chairman of NBAC, on December 6, 2000 (Public Citizen, 2000).

pacity to review and conduct clinical trials, (2) involving community representatives in the design and implementation of clinical trials in developing countries, and (3) sponsoring workshops at which international researchers can share their knowledge about the informed consent process.

The UK's Nuffield Council on Bioethics discussed NBAC's work on clinical trials research in developing countries at some of its meetings, and presentations on studies commissioned by NBAC as part of its report on international research were made at the first and second meetings of the Global Forum for Bioethics in Research. Details on these two international responses to the NBAC *Ethical and Policy Issues in International Research* report are provided next.

Response from the United Kingdom

NBAC's work and its draft report on clinical trials research in developing countries were mentioned at meetings of the Nuffield Council on Bioethics and at meetings of the council's Working Party on the Ethics of Healthcare-Related Research in Developing Countries (Nuffield Council on Bioethics, 2000a and 2001). At one of the Nuffield Council meetings, it was suggested that there was the possibility of teaming with NBAC to hold Working Party meetings in countries besides the UK and United States.

Response from the Global Forum for Bioethics in Research

The Global Forum for Bioethics in Research brings together individuals involved in medical research in developing countries to share their views with each other and with organizations that support clinical research. The Global Forum is an ongoing enterprise designed to examine the challenges that arise from cross-cultural research, and to provide guidance for improving institutional capabilities in bioethics in developing countries.

At the first Global Forum in November 1999 in Bethesda, Maryland, which was cosponsored by the NIH, WHO, and the Pan American Health Organization, Dr. Jeremy Sugarman spoke about his study that was commissioned by NABC on international perspectives on protecting human research participants. At the second meeting in

October 2000 in Bangkok, Thailand, which was cosponsored by WHO, NIH, the United Kingdom's Medical Research Council, the South African Medical Research Council, and other international agencies, Dr. Adnan Hyder spoke about a survey commissioned by NBAC that examined the attitudes and experiences of investigators from developing countries regarding human subjects regulations in the United States. At both meetings, the majority of those in attendance were from developing countries.

RESPONSE TO THE *ETHICAL AND POLICY ISSUES IN RESEARCH INVOLVING HUMAN PARTICIPANTS* REPORT

While most research involving human participants has been conducted ethically and has resulted in great benefits to society, some regrettable abuses involving research participants have occurred in the past. A number of well-publicized studies, such as the Tuskegee Study of Untreated Syphilis in the Negro Male,[1] experiments with hepatitis at the Willowbrook State School for the Retarded,[2] and federally sponsored studies of ionizing radiation on humans,[3] have led to a certain amount of distrust toward researchers.

On May 16, 1997, President Clinton, on behalf of the United States, formally apologized to the participants in the Tuskegee Syphilis Study and their families, stating that "all our people must be assured that their rights and dignity will be respected as new drugs, treatments, and therapies are tested and used" ("Remarks by the President . . .," 1997). The next day, at an NBAC meeting in Arlington, Virginia, NBAC unanimously resolved, "No person in the United States

---

[1]The study, conducted by the U.S. Public Health Service to examine the natural course of untreated syphilis in African-American men, began in 1932 and continued until 1972, almost 30 years after penicillin was discovered as a cure for syphilis. The subjects were all impoverished sharecroppers from Tuskegee in Macon County, Alabama. They were unknowing participants in the study, they were not told that they had syphilis, nor were they offered effective treatment.

[2]These experiments, which were conducted during the 1950s at the Willowbrook School in New York, involved injecting a mild strain of the hepatitis virus into children at the time of admission to the school. Although parental consent was obtained, questions were raised about whether that consent was truly voluntary.

[3]Most of these experiments were conducted between 1944 and 1974 in numerous places by the Department of Energy (in more than 400 human radiation experiments), the Department of Defense, and several other agencies.

should be enrolled in research without the twin protections of informed consent by an authorized person and independent review of
the risks and benefits of the research."

Many of the components of today's system of oversight for the protection of research participants arose in response to ethical lapses
and abuses in the research system. The concept of informed consent
was originally established by the Nuremberg Code in response to
atrocities committed by Nazi physicians in World War II. The first
principle of the Nuremberg Code states that "the voluntary consent
of the human subjects is absolutely essential" (United States v. Karl
Brandt et al., 1949).

The requirement for independent review of research by a committee
composed of the investigators' "institutional associates" (i.e., an
Institutional Review Board) was established by the Public Health
Service (PHS) in its "Policy for Clinical Investigators with Human
Subjects" (Public Health Service, 1966, pp. 475–476). The PHS policy
was implemented in response to newspaper accounts of an NIH-
funded study in the early 1960s at the Brooklyn Jewish Chronic
Disease Hospital in which investigators had injected cancerous cells
into elderly patients.[4]

The *Belmont Report,* written by the National Commission for the
Protection of Human Subjects of Biomedical and Behavioral Research (National Commission), identified three fundamental ethical
principles applicable to research with human participants—respect
for persons, beneficence,[5] and justice—which, respectively, provide
the ethical rationale for the requirements of informed consent,
assessment of risk and potential benefits, and selection of research
participants (National Commission . . . , 1979).

---

[4]This case came to light in 1964 when newspapers began to describe the study. In
1965, the board of directors of the Jewish Chronic Disease Hospital brought a lawsuit
against the hospital to have access to records of 22 patients who received injections of
suspensions of cultures of human cancer tissue. The investigators claimed to have obtained informed consent from the patients; however, many of the patients were incapacitated or did not speak English, and those who were able to give consent were not
told that the cells were cancerous.

[5]The National Commission used the term "beneficence" to mean the obligation to do
no harm and to maximize possible benefits.

Recommendations in the first and second biennial reports of the President's Commission for the Study of Ethical Problems in Medicine and Biomedical and Behavioral Research (President's Commission) established the basis of the Common Rule, the core component of the federal research oversight system (President's Commission . . ., 1981 and 1983). The Common Rule applies to all research involving human participants "conducted, supported or otherwise subject to regulation by any federal department or agency which takes appropriate administrative action to make this policy applicable to such research."

Executive Order 12975 of October 3, 1995, which established NBAC, required that each federal department and agency "that conducts, supports, or regulates research involving human subjects shall promptly review the protections of the rights and welfare of human research subjects that are afforded by the department's or agency's existing policies and procedures" and report the results of their review to NBAC. In 1996, federal departments and agencies began responding to NBAC's request for information by reporting on their current protections, and by evaluating those protections and taking steps to strengthen them.

Based on the agency reports and its own examination of the issues, NBAC concluded that the Common Rule "has significantly reduced, but not eliminated, the possibility for harm" to research participants. Consequently, in May 1999, NBAC wrote to President Clinton indicating several areas of concern with the current federal system for the protection of human research participants. Of primary concern to NBAC was the finding that federal protections for people serving as participants in research did not yet extend to all Americans. In particular, there are significant sectors of privately funded research that are not governed by either state or federal law. In addition, NBAC found that the interpretation and implementation of the Common Rule was confusing or unnecessarily burdensome for many federal agencies.

In October 1999, the White House asked NBAC to undertake a thorough examination of the federal system of protections for human research participants. Specifically, NBAC was asked to assess the adequacy of the current federal system of protections, review relevant statutes and regulations with particular attention to the effectiveness of the Common Rule and its applicability to the full range of government-sponsored research activities involving human research participants, and examine the strengths and weaknesses of the infrastructure responsible for ensuring the entire system's integrity.

On December 19, 2000, NBAC released a draft report on research involving human participants for public comment and received 214 responses. In May 2001, NBAC finalized its recommendations and on August 20, 2001, issued its final report on the subject, *Ethical and Policy Issues in Research Involving Human Participants.* The report contained 30 recommendations proposing a new oversight system that would protect all research participants while encouraging ethically sound research and supporting the continued advancement of science.

Specifically, NBAC concluded that the federal oversight system should protect the rights and welfare of human research participants, regardless of whether the research is publicly or privately funded, and further recommended that legislation should be enacted to provide such protection. NBAC recommended that there be a unified and comprehensive federal policy embodied in a single set of regulations and guidance, and called for legislation to create a single, independent federal office responsible for the oversight of human research participants. Other recommendations addressed verification of compliance, certification and accreditation, conflicts of interest, IRB membership, informed consent, continuing review, adverse event reporting, review of minimal risk research, and compensation for research-related injuries.

GOVERNMENT RESPONSE

Of the 30 recommendations that NBAC made in *Ethical and Policy Issues in Research Involving Human Participants,* 24 were directed to Congress or the federal government, or recommended changes to federal policy. NBAC specifically called for federal legislation to be enacted to extend the federal oversight of human research to protect participants in all research, whether publicly or privately funded, and to create a new federal office to oversee the protection of all research participants (Recommendations 2.1 and 2.2, respectively). NBAC also recommended that a unified, comprehensive federal policy embodied in a single set of regulations and guidance be created to govern all types of research involving human participants (Recommendation 2.3), and made 16 other recommendations that laid out specific changes that needed to be made in the federal policy for the protection of human research participants.

NBAC's examination of the oversight system for the protection of people serving as participants in research received attention from the Congress, DHHS, NIH, and the Veterans Health Administration (VHA). Legislation was introduced in Congress that was consistent with NBAC's recommendation for one set of regulations governing research involving human participants, and included a permanent authorization for NBAC. In addition, NBAC commissioners and others discussed NBAC's work on research involving human research participants several times in congressional testimony. References to NBAC's analysis of the relocation of OPRR from NIH to the Office of the Secretary at DHHS were made in a 1999 report (discussed later in this chapter) to the NIH Advisory Committee to the Director (ACD) and by the acting NIH director at an Association of American Medical Colleges annual meeting (also discussed later). Federal legislation, congressional testimony, guidelines, policy statements, and other such documents that refer to NBAC's work on research involving human participants are detailed next.

Response from Congress

Legislative Action. The Human Research Subject Protections Act of 2000, H.R. 4605, was introduced on June 8, 2000, by Representative Diana DeGette (D-Colo.). The bill was referred to the House Commerce Committee and had 11 cosponsors, including Representative Henry Waxman (D-Calif.) and Representative John Mica (R-Fla.). The bill applied the Common Rule to all research regardless of the funding source, and established a single oversight office for the protection of human research participants within DHHS. This bill was consistent with NBAC's recommendations for a single set of regulations governing research involving human participants that is promulgated and interpreted by a single independent federal office. The bill cited NBAC's findings that federal regulations do not contain adequate protections for certain vulnerable populations and do not adequately address the research use of human biological materials. The bill also included a permanent authorization for NBAC. The bill was endorsed by the American Association of Medical Colleges, the National Organization for Rare Diseases, and the Citizens for Responsible Care & Research.

At a September 26 markup (when proposed legislation is debated and amended) of H.R. 1798, the "Clinical Research Enhancement Act," Representative DeGette offered H.R. 4605 as an amendment. Objections were raised to the amendment, and the amendment was withdrawn. On July 25, 2001, DeGette and Representative Jim Greenwood (R-Pa.) announced their intention to introduce an updated version of H.R. 4605 in September of that year. A "Dear Colleague" letter circulated by Representatives DeGette and Greenwood stated that the updated bill adopts several of NBAC's recommendations, including the establishment of regulatory oversight commensurate with the level of research risk, strengthening the conflicts of interest provisions, tightening monitoring requirements of ongoing research projects, and strengthening the IRB system.

Congressional Testimony. A May 3, 2000, hearing by the House Subcommittee on Criminal Justice, Drug Policy, and Human Resources focused on an April 2000 report from Office of the Inspector General (OIG), DHHS. The report, entitled *Protecting Human Research Subjects: Status of Recommendations,* detailed the status of human research recommendations made by the OIG in 1998 (see the subsection below describing the response from the DHHS). George Grob, OIG deputy inspector general for Evaluations and Inspections, and Dr. William Raub, deputy assistant secretary for Science Policy, DHHS, testified at the hearing. Both referred to NBAC's *Ethical and Policy Issues in Research Involving Human Participants* report in their written statements but not in their oral remarks. Grob reviewed the findings of the IOG's April 2000 report and indicated that while there had been a substantial increase in the enforcement of federal protections of research participants, few of the recommended reforms in the 1998 report had been enacted. Raub testified that the contents of the OIG report created some concern for DHHS, but he also noted that the report did not reveal much evidence of actual harm to research participants.

The House Subcommittee on Oversight and Investigations of the House Committee on Veterans Affairs had a hearing on medical research involving human participants at the Department of Veterans

Affairs on September 28, 2000. Greg Koski, M.D., director of OHRP, testified that the report by the OIG in 1998 raised many basic issues of protection of human participants, and he said that OHRP endorsed the report and would act on its recommendations. Dr. Koski proposed that there should be protections in place for all research involving human participants, including certification of individual investigators, accreditation of IRBs, and issuance of assurances. There was no mention of NBAC in his oral presentation, although his written statement noted NBAC's *Ethical and Policy Issues in Research Involving Human Participants* report. His statement noted that the Institute of Medicine would be asked to conduct a study of the system for protection of human participants in research to determine the extent to which the system addresses issues raised by the OIG and by the recommendations in NBAC's report.

Eric Meslin, then NBAC executive director, testified before the Subcommittee on Human Resources of the Committee on Government Reform and Oversight, U.S. House of Representatives, on June 11, 1998. Dr. Meslin testified about NBAC's work on issues in research involving human participants, including research involving persons with mental disorders that may affect their decisionmaking capacity and research involving human biological materials. He also mentioned that NBAC would address issues in international research, specifically the rules that ought to apply when the United States conducts or supports research in other countries. Finally, he described NBAC's planned comprehensive assessment of the federal system of oversight of protections for human research participants, indicating that, at the time, NBAC had begun its work on this subject by conducting a survey of federal agencies to determine their implementation of the Common Rule.

In the capacity of president of The Hastings Center, a research institute that focuses on issues in health care, biotechnology, and the environment, NBAC commissioner Thomas Murray testified at the Senate Public Health Subcommittee's second hearing on gene transfer research on May 25, 2000. Murray discussed the work pressures on IRBs, conflicts of interest in clinical research, and the need for coordination between NIH and FDA for the review of gene transfer research and adverse event reporting.[6] In responding to questions, Murray described NBAC's report on oversight of research involving human participants and discussed NBAC's proposed model of designating one federal office with primary authority over research involving human participants. He also proposed that there be increased research into the informed consent process and what participants actually learn from it. Dr. Raub also testified at the hearing and mentioned the NBAC report on research involving human participants during the hearing's question-and-answer period.

On October 16, 2001, NBAC commissioner Alexander Capron testified on behalf of NBAC about the commission's *Ethical and Policy Issues in Research Involving Human Participants* report before the Senate Public Health Subcommittee. Capron discussed several of NBAC's recommendations and highlighted two as being the most important: (1) the recommendation for a unified and comprehensive federal policy "embodied in a single set of regulations and guidance" that would apply to all types of research involving human participants, and (2) the recommendation for legislation to create a single, independent federal office responsible for the oversight of human research participants.

Response from the Department of Health and Human Services

Following President Clinton's formal apology on behalf of the United States to the participants of the Tuskegee Syphilis Study and their families, he directed DHHS "to identify strategies to improve the participation of communities, especially minority communities, in research and to build trust between researchers and communities." *Building Community Partnerships in Research,* prepared by a steering committee composed of the CDC, NIH, FDA, HRSA, IHS, and Substance Abuse and Mental Health Services Administration, is DHHS's response to the president's request (Centers for Disease

---

[6]While NBAC did not prepare a report specifically on gene transfer research, it did hear testimony at the March 1, 2000, NBAC commissioners' meeting from the NIH and FDA on how those agencies have dealt with reported problems associated with gene transfer clinical trials. The testimony was requested by NBAC as part of its deliberations on the oversight system for the protection of research participants.

Control and Prevention et al., 1998). The purpose of the steering committee report is to provide "a framework through which Federal health agencies can establish an ethical basis for community-based research, enhance scientific and public credibility, and provide mechanisms to help build trust in health research." The report mentions that NBAC was created to review current regulations, policies, and procedures that ensure the protection of research participants.

In June 1998, the OIG issued a report entitled *Institutional Review Boards: A Time for Reform* that warned that the effectiveness of IRBs was in jeopardy and called for widespread reform of IRBs (Office of the Inspector General, 1998). In April 2000, the OIG issued *Protecting Human Research Subjects: Status of Recommendation*, which provided an update of NIH and FDA responses to recommendations in the June 1998 OIG report (Office of the Inspector General, 2000). This update mentioned NBAC's continued work on protections for human research participants and urged OHRP to pay close attention to NBAC's forthcoming recommendations. In the April 2000 report, the OIG found that while there had been a "substantial increase in the enforcement of Federal human-subject protection requirements," few of the recommended reforms detailed in the June 1998 report had been enacted. Furthermore, many of the OIG's recommendations to DHHS called for changes in the Common Rule, but because any change to the Common Rule requires the concurrence of all 17 federal agencies that are signatories to it, the OIG concluded that timely implementation of these recommendations was not possible.

Response from the Centers for Disease Control and Prevention

In a news update issued on February 7, 2001, the CDC National Center for HIV, STD (sexually transmitted disease), and TB (tuberculosis) Prevention (NCHSTP) issued a memo alerting its partners to the release of NBAC's draft report on oversight of research involving human participants ("NBAC Draft Oversight Report/ . . .," 2001). The "partners" are those with whom the NCHSTP collaborates in research, those who represent groups who may participate in NCHSTP research, or those who may otherwise be interested in the protection of research participants. The memo summarized NBAC's key recommendations. It also mentioned that NBAC's report "comes at a time of increased Congressional scrutiny of and public concern over research practices," and that "NBAC's recommendations are likely to influence Congressional response on this topic."

Response from the National Institutes of Health

A June 3, 1999, report to the NIH ACD from the Office for Protection from Research Risks Review Panel recommended that the OPRR should be administratively relocated from within NIH to the Office of the Secretary of HHS (Office for Protection from Research Risks, 1999). In a commentary accompanying the OPRR Review Panel's recommendation to relocate OPRR from within NIH, the review panel explains that commissioned papers prepared for NBAC had previously found that, because it is part of the NIH and because it reports to supervisors within the NIH, the OPRR was not perceived as an independent office and this situation raised concerns about conflicts of interest. The review panel concluded that "relocating OPRR was the only way to address these perceptions and concerns and to ensure OPRR's independence and maximize its effectiveness." HHS Secretary Donna Shalala accepted the recommendations of the panel and decided that the office would report to the Assistant Secretary of Health. DHHS officials also decided to assign the OPRR duties related to the protection of human research participants to a new entity called the Office for Human Research Protections (OHRP) while keeping the animal protection duties at the NIH. (In May 2000, DHHS announced a reorganization of OPRR, and in June 2000, OPRR became OHRP.)

In a speech entitled "NIH and the Academic Research Community: Partnership for Clinical Trials" at the Association of American Medical Colleges Grand Annual Meeting on May 7, 2000, Ruth Kirschstein, M.D., then acting director of the NIH, discussed the elevation of OPRR from the NIH to the Office of Public Health and Science within the Office of the Secretary at DHHS and its new name, the Office for Human Research Protections (Kirschstein, 2000). Dr. Kirschstein mentioned that this action was recommended in a study by the NIH ACD and in an independent analysis by NBAC.

NBAC's work on the oversight of research involving human participants was mentioned at several COPR meetings and a meeting of the Human Subjects Research Advisory Committee (HSRAC) to OHSR at

NIH. At the April 6–7, 2000, COPR meeting, Gary Ellis, then director of OPRR, discussed the protection of human research participants and highlighted NBAC's 1997 recommendation for universal protections for research participants, regardless of whether the research was federally funded or regulated by FDA (NIH Director's Council . . ., 2000a). At the October 31-November 1, 2000, COPR meeting, Thomas Murray, then an NBAC Commissioner, provided an update on NBAC's work on the oversight of research involving human participants (NIH Director's Council . . ., 2000b).

At the May 1, 2001, COPR meeting, the COPR Human Research Protections Working Group reviewed federal administrative activities and programs throughout HHS that are relevant to the protection of human research participants, and highlighted NBAC's upcoming report on the subject (NIH Director's Council . . ., 2001). NBAC's draft recommendations for a national system of oversight for the protection of human research participants and for the composition of IRB members were mentioned during a discussion about OHRP's new guidelines entitled "Federalwide Assurance of Protection for Human Subjects" at the January 12, 2001, HSRAC meeting ("Assurance of Protection . . .," 2001).

Response from the Veterans Health Administration

The Veterans Health Administration National Center for Ethics is the VHA's primary office for addressing ethical issues associated with patient care, health care management, and research. An article entitled "National Bioethics Advisory Commission (NBAC) Scrutinizes Current System for Protecting Research Subjects" in the "National Ethics News" of the Winter 2001 edition of *news@vhaethics*, the National Center for Ethics newsletter, described NBAC's draft report and recommendations on the oversight of research involving human participants ("National Bioethics Advisory Commission . . .," 2001). The article also notified readers that NBAC was seeking public comments on the draft report and directed them to NBAC's Web site for further information.

RESPONSE FROM PROFESSIONAL SOCIETIES, ORGANIZATIONS, AND FOUNDATIONS

In its report *Ethical and Policy Issues in Research Involving Human Participants*, NBAC recognized that professional organizations have an important role to play in the protection of human research participants, especially in the area of research ethics education. NBAC made recommendations that direct the federal government to cooperate with professional societies on education related to protecting human research participants, and also to cooperate with professional societies on discussions and research on emerging human research protection issues (Recommendations 3.2 and 7.2, respectively). NBAC concluded that "it is time to encourage and foster the participation of groups often overlooked in the protection of research participants, including professional organizations, journal editors, and patient advocacy groups."

The Alliance for Human Research Protection (AHRP) supported NBAC's recommendations on risk assessment in a letter to the FDA about safeguards for children in clinical trials, which is discussed next. The AAAS, AMA, and AAMC have mentioned NBAC's work on the protection of human research participants in their newsletters. Details about the responses of these organizations to the *Ethical and Policy Issues in Research Involving Human Participants* report follow.

Response from the Alliance for Human Research Protection

AHRP commented on the FDA Interim Rule "Additional Safeguards for Children in Clinical Investigations of FDA-Regulated Products" in an August 6, 2001, letter to the acting commissioner of FDA (Alliance for Human Research Protection, 2001). AHRP supported the FDA's decision to bring FDA regulations into compliance with the Children's Health Act of 2000 by adopting Subpart D of 45 CFR 46, which provides additional protections for children involved in research. AHRP also supported the FDA's exclusion of Section 46.408 (c), which allows waiver of parental or guardian permission by IRBs under certain circumstances, based on the conclusion that the waiver is not permitted under FDA law. In its comments to the FDA, AHRP referred to NBAC's *Ethical and Policy Issues in Research Involving Human Participants* report. Specifically, AHRP stated that NBAC's recommendations on risk assessment would "improve research safe-guards for adults and children," and that NBAC's recommended framework for analyzing risk was "scientifically sound." AHRP also stated that NBAC's recommendation for review of very high-risk re-search by a national review panel "serves the public interest and should be adopted."

Response from the American Association for the Advancement of Science

AAAS Report XXVI: Research and Development FY 2002 mentions some of NBAC's recommendations on the oversight of research involving human participants. The report's chapter on "Political and Policy Context for the FY 2002 Budget" (Nelson and Cooper, 2001) predicted that Congress and the administration may decide to change the way the federal government oversees the protection of participants in research after the publicity surrounding the 1999 death of a participant in a gene transfer clinical trial. The chapter also highlights two of NBAC's main recommendations—specifically, to extend protections to research participants in all research, whether federally or privately funded, and to establish a single, independent office for the oversight of research involving human participants.

In June 2001, the Center for Science, Technology, and Congress at AAAS published an article in its newsletter, *Science & Technology in Congress*, about the publication of NBAC's recommendations on the protection of human research participants ("NBAC Proposes . . .," 2001). The article highlighted NBAC's key recommendation that the federal oversight system should protect all research participants, re-gardless of whether the research is publicly or privately funded, and that federal legislation should be enacted to provide such protection. The article also discussed several of NBAC's other recommendations, including enacting a comprehensive federal policy for protecting re-search participants, establishing a single, independent oversight of-fice, improving IRB review, and managing conflicts of interest.

Response from the American Medical Association

Two articles in the AMA's *American Medical News* mentioned NBAC's work on the oversight of research involving human participants. The first article, in the July 10/17, 2000, issue (Foubister, 2000), discussed two of the ethical concerns associated with clinical trials— a physician serving as both the doctor of a patient and as an investigator of a clinical trial involving that patient, and sponsors of clinical trials making payments to physicians for enrolling patients in clinical trials. The article mentioned that NBAC planned to release a report on the protection of human research participants by the end of that year. The second article, in the February 19, 2001, issue, focused specifically on NBAC's draft report on the oversight of research involving human participants (Foubister, 2001). The article quoted several individuals who had criticized the report and indicated that there was little support for NBAC's recommendations on creating a new independent federal agency to promulgate a new set of regulations to govern this research.

Response from the Association of American Medical Colleges

In the *AAMC Washington Highlights* weekly news bulletin, the AAMC reported on two of NBAC's meetings on research involving human participants. The April 14, 2000, issue summarized NBAC's April 6–7, 2000, meeting during which the expansion of IRB regulations was discussed ("NBAC Considers . . .," 2000). The July 21, 2000, issue summarized NBAC's July 10–11, 2000, meeting during which NBAC heard from six panels that discussed their perspectives on research involving human participants ("NBAC Continues . . .," 2000).

INTERNATIONAL RESPONSE

Although the NBAC report *Ethical and Policy Issues in Research Involving Human Participants* strictly addresses the oversight system for the protection of research participants in the United States, its policy recommendations may be relevant to the protection of research participants in other countries. In this regard, it is interesting to note that Canada's Centre on Governance mentioned NBAC's work in its study of Canada's system of governance for research involving humans.

Canada

The Governance of Health Research Involving Human Subjects, a report of the Centre on Governance of the University of Ottawa (Centre on Governance, 2000), was written in response to a request from the Law Commission of Canada for the center to conduct a study of the current governance process for research involving humans, as established by the "Tri-Council Policy Statement: Ethical Conduct for Research Involving Humans."[7] The report mentions NBAC within the context of an historical background on the protection of research participants in the United States. The Centre on Governance recommended that the Tri-Council Policy Statement should become the "gold standard" for all research in Canada, governing not only research conducted by universities, but also research performed by the government, nonprofit groups, and the private sector.

---

[7]The Tri-Council Policy Statement is a statement from three major research granting agencies in Canada—the Medical Research Council, the Natural Science and Engineering Council, and the Social Sciences and Humanities Research Council.

SUMMARY

The National Bioethics Advisory Commission was established by President Clinton in October 1995, and the commission met for the first time one year later on October 4, 1996. Its charter expired on October 3, 2001. During its tenure, NBAC submitted six major reports to the president, which collectively contained 120 recommendations on improving the protection of human research participants while supporting the continued advancement of science and promoting ethically sound research. These reports and the recommendations they contain have informed the United States and foreign governments, international groups, the research community, and the public about several complex bioethical issues. Those issues include cloning of human beings, research involving persons with mental disorders that may affect decisionmaking capacity, research with human biological materials, embryonic stem cell research, U.S.-sponsored clinical trials in developing countries, and the protection of all human participants in research.

NBAC's contribution to the public policy process was multifaceted. Each of NBAC's reports synthesized a substantial body of information and provided analysis of some very important and quite contentious issues. NBAC's in-depth examination of the scientific, legal, and ethical questions surrounding each issue it addressed led to the formulation of reasoned recommendations for action on these issues. And while not everyone agreed with the recommendations that NBAC made, the commission's reports did provide useful information and clarification of pertinent issues for those who may not have agreed with the commission's conclusions.

Responses to NBAC's work varied widely, from a passing mention of NBAC or its work to thoughtful consideration of NBAC's analyses and recommendations, and from partial or complete acceptance of the recommendations and their incorporation into policy to outright rejection of them. However, because much of NBAC's contribution to the policymaking process was to inform public discussion and debate of the issues it addressed, acceptance of NBAC's recommendations is not a full measure of the commission's contribution to public policy. In addition, this aspect of NBAC's contribution to the policymaking process is not easily measured.

There is ample evidence that NBAC's work has stimulated discussion and has informed the public policy debate on bioethical issues in this country and elsewhere. As of this writing, no federal or state legislation has been passed that is based on any of NBAC's recommendations. However, the government agencies responsible for a major portion of federally funded research involving human participants—the NIH, FDA, and CDC—have adopted several of NBAC's recommendations and have issued guidelines for researchers to follow based on those recommendations. There is also evidence that professional societies have taken note of NBAC's recommendations—they have released policy statements, issued guidance, and developed educational materials for their members to consider based on NBAC's reports and recommendations. Finally, it is clear that other countries and international organizations have been informed by NBAC's reports and recommendations, given that they have cited the commission's work in various publications and have endorsed some of its recommendations.

Clearly, NBAC has increased the awareness of U.S. and foreign governments, international groups, the research community, and the public about complex bioethical issues, thereby helping to provide a forum for public debate of those issues. Furthermore, recommendations stemming from NBAC's work have been incorporated into the U.S. system of oversight for the protection of human research participants.

The two NBAC reports that received the most attention were *Cloning Human Beings* and *Ethical Issues in Human Stem Cell Research*. These two reports prompted responses from the President of the United States, Congress, federal agencies, professional societies, other countries, and international organizations. The topics of these reports, cloning and stem cell research, are highly controversial areas of science that have been extensively covered in the media and continue to be debated today by the President, Congress, the President's Council on Bioethics, and many others.

Although the response to the NBAC reports *Research Involving Persons with Mental Disorders That May Affect Decisionmaking Capacity* and *Research Involving Human Biological Materials* was not as extensive as the response to the NBAC reports on cloning and human stem cell research, they nevertheless received a fair amount of attention, especially from the NIH. Both of the reports also prompted responses from Congress and professional societies. It is interesting to note that both reports also prompted responses from state governments. And while *Research Involving Human Biological Materials* did receive some attention from the international community, no mention of the *Research Involving Persons with Mental Disorders That May Affect Decisionmaking Capacity* report was found in the regulations, guidance, policy statements, or documents of other countries or international organizations.

The response to NBAC's last two reports, *Ethical and Policy Issues in International Research: Clinical Trials in Developing Countries* and *Ethical and Policy Issues in Research Involving Human Participants*, was more limited than the response to the other four reports issued by NBAC. This may be due, in part, to their late release dates, which did not provide adequate time for thorough consideration by interested parties. In addition, these reports were issued shortly before the expiration of NBAC's charter and the establishment of President Bush's new President's Council on Bioethics, which may have attenuated the response to these two reports.

The NBAC report *Ethical and Policy Issues in International Research* by far received the least response from Congress, federal agencies, professional societies, other countries, and international organizations. However, NBAC's last report, *Ethical and Policy Issues in Research Involving Human Participants*, received more attention than would be expected given that it was released just six weeks prior to the expiration of the commission.

The greater-than-expected response to *Ethical and Policy Issues in Research Involving Human Participants* can be attributed to several factors:

- First, NBAC's charter laid the foundation for that report, stating "[a]s a first priority, NBAC shall direct its attention to consideration of protection of the rights and welfare of human research subjects." The other five reports that NBAC issued focused on different aspects of research involving human participants.

- Second, much of the attention the report received can be attributed to the fact that NBAC started preliminary work on the report soon after the commission was convened in the fall of 1996. One of the first things that NBAC did was to request that all relevant federal agencies report to NBAC on their current protections for human research participants. A letter conveying NBAC's preliminary findings about the federal system for protecting human research participants was sent to the president in May 1999.

- Third, in October 1999, Dr. Neal Lane, then assistant to the president for science and technology, formally requested that NBAC examine the adequacy of the current system for the protection of human research participants.

- Finally, NBAC commissioners and staff increased the visibility of the *Ethical and Policy Issues in Research Involving Human Participants* report by informing legislators, key agency personnel, and relevant professional societies about its recommendations long before the final report was issued.

Not all of the responses to NBAC's reports and recommendations were favorable. In fact, some of the reports were highly criticized for their narrow scope and for their recommendations that called for a cumbersome regulatory framework. And some of NBAC's recommendations raised concerns that they were impractical and would impede research, while other recommendations were criticized for not being restrictive enough.

Whether or not a commission's recommendations are implemented depends in part on how they are formulated and to whom they are targeted. NBAC took this into account when it was formulating its recommendations, carefully considering the types of policy changes it recommended (e.g., legislation, regulation, guidance, education) and to whom they were directed (e.g., Congress, federal agencies, professional societies, the private sector). For example:

- In both *Cloning Human Beings* and *Ethical and Policy Issues in Research Involving Human Participants,* NBAC recommended that Congress should enact legislation.

- In *Research Involving Persons with Mental Disorders That May Affect Decisionmaking Capacity,* NBAC recommended that the states should enact legislation.

- In *Cloning Human Beings,* NBAC recommended that the U.S. government should cooperate with other nations and international organizations on their respective policies regarding cloning of human beings.

- In *Ethical and Policy Issues in International Research: Clinical Trials in Developing Countries,* NBAC recommended that the FDA should not accept data obtained from research that does not provide the substantive ethical protections outlined in the report.

- In *Ethical and Policy Issues in Research Involving Human Participants,* NBAC recommended that the federal government issue guidance on both research with vulnerable populations and informed consent.

- In *Research Involving Persons with Mental Disorders That May Affect Decisionmaking Capacity, Research Involving Human Biological Materials: Ethical Issues and Policy Guidance,* and *Ethical and Policy Issues in Research Involving Human Participants,* NBAC recommended that professional societies should develop ethics training for investigators.

- In *Cloning Human Beings,* NBAC recommended that federal agencies should support public education on the ethical and social implications of biomedical research.

- In *Cloning Human Beings* and *Ethical Issues in Human Stem Cell Research,* NBAC recommended that the private sector voluntarily comply with NBAC's recommendations and the ethical principles underlying them.

As mentioned earlier, the response to NBAC's six reports has varied from a passing mention of the commission's work to the issuance of guidance by federal agencies. Although no federal or state legislation has been signed into law on the basis of NBAC's recommendations, federal and state bills based on the recommendations have been introduced, and Congress was informed about NBAC's work through congressional testimony by NBAC commissioners and other individuals. Professional societies have released policy statements, issued guidance, and developed educational materials based on NBAC's reports and recommendations, and other countries and international organizations have endorsed some of NBAC's recommendations.

During NABC's tenure, 13 bills were introduced in Congress that mentioned NABC.[1] Four of those bills were introduced in response to the cloning debate, eight dealt with the privacy of genetic and medical information, and one addressed the system for oversight for the protection of human research participants. Three of the 13 bills were based directly on NBAC's recommendations and would have authorized a continuation of NBAC's tenure. Three bills directed the secretary of HHS to consider the findings of NBAC when formulating recommendations on the privacy of health information. The rest of the bills called for studies to be performed or recommendations to be formulated by NBAC on various topics. In addition, bills on human cloning, decisionmaking capacity, and genetic privacy that mentioned NBAC or were based on NBAC's recommendations were introduced in the legislatures of four states. To date, none of these bills introduced at either the state or federal level has been signed into law.

In addition, NBAC commissioners and staff testified before Congress 18 times to discuss both ongoing work and completed projects on various issues, including cloning, decisionmaking capacity, human stem cell research, and the protection of human research participants. Furthermore, others invited to testify before Congress referred to NBAC's reports and recommendations, including members of Congress, officials from DHHS, and representatives of professional societies.

---

[1]Appropriations bills in the 104th Congress (S.J. RES. 63 and H.R. 4278) and 105th Congress (H.R. 2378, S. 2312, and H.R. 4104) that authorized "interagency financing" of NBAC are not included in the number of bills that mention NBAC.

Federal agencies have issued guidance and policy statements in response to NBAC's reports and recommendations. The NIH, the government agency responsible for the majority of federally funded research involving human participants, has issued guidance based on recommendations made in the *Research Involving Persons with Mental Disorders That May Affect Decisionmaking Capacity, Research Involving Human Biological Materials,* and *Ethical Issues in Human Stem Cell Research* reports. In addition to issuing guidance, the NIH also created a new psychiatric research review panel and announced its intent to sponsor research on assessing decisionmaking capacity; both actions were consistent with NBAC's recommendations. The FDA has also adopted NBAC's terminology and recommendations on informed consent found in *Research Involving Human Biological Materials,* and NIOSH instructed its investigators who perform research involving persons with developmental disabilities to review NBAC's recommendations on informed consent in *Research Involving Persons with Mental Disorders That May Affect Decisionmaking Capacity.*

Professional societies have also taken note of NBAC's recommendations and have released policy statements, issued guidance, and developed educational materials based on NBAC's reports and recommendations for their members to consider. For example, several professional and scientific societies, and organizations representing the biotechnology and pharmaceutical industries, called for a five-year voluntary moratorium on human cloning, consistent with NBAC's recommendation in *Cloning Human Beings.* The Alzheimer's Association funded research on the potential impact of NBAC's recommendations in *Research Involving Persons with Mental Disorders That May Affect Decisionmaking Capacity.* Intermountain Health Care issued policies and procedures for IRBs to follow in reviewing research involving human biological materials, and the Online Ethics Center for Engineering and Science developed an on-line teaching module, all of which were based on NBAC's recommendations in *Research Involving Human Biological Materials.* The Endocrine Society in its Code of Ethics and the AMA in its official policy supported NBAC's report and recommendations on *Ethical Issues in Human Stem Cell Research.* Although position statements, policies, and guidance from medical and scientific organizations lack the force of federal regulations, they nevertheless can be influential in shaping the behavior and practices of the scientific and medical communities.

Other countries and international organizations have been informed by NBAC's reports and recommendations and have supported some of NBAC's recommendations. Several countries and international organizations have also cited NBAC's work in various reports and publications. In addition, the NBAC reports have been widely circulated internationally.

It should also be noted that while some policy recommendations are implemented fairly quickly, many years might pass before others are implemented, and many are never implemented. For example, the response to the National Commission's April 1979 *Belmont Report* occurred fairly quickly (National Commission . . ., 1979). The *Belmont Report*, a seminal report on research with human participants, identified three fundamental ethical principles applicable to research with human participants—respect for persons, beneficence, and justice. Application of these three principles to research provided the ethical rationale for the requirements for informed consent, assessment of risk and potential benefits, and selection of research participants, respectively. In response to the Belmont Report, the DHHS and FDA simultaneously revised their regulations, which were signed by the secretary of HHS in January 1981.

In contrast, the response to the 1981 President's Commission report entitled *Protecting Human Subjects: The Adequacy and Uniformity of Federal Rules and Their Implementation* took quite a bit longer (President's Commission, 1981). The President's Commission recommended that uniform regulations be implemented across all federal agencies for the protection of human research participants. However, the Federal Policy for the Protection of Human Subjects (the Common Rule), which provided some standardization of regulations across federal agencies, was not codified until 1991, ten years after the initial recommendations of the President's Commission. Furthermore, neither the National Commission's recommendations regarding research involving individuals institutionalized as mentally infirm nor the President's Commission's recommendations on compensating individuals for research-related injuries were ever implemented. Therefore, the impact of NBAC's reports and recommendations on the policymaking process should be assessed over time to determine both the immediate and long-term responses to the commission's recommendations.

In addition to legislation, regulations, guidance, and position statements, NBAC has contributed to the policymaking process in several other ways. For example, in January 1998, NBAC launched its Web site as a comprehensive source of information on the commission and as a central location for posting its reports, meeting agendas, meeting transcripts, and other such materials. Many government agencies, universities, and national and international groups established links on their Web sites to the NBAC site and relied on it for up-to-date information about bioethics developments in the United States. In addition, NBAC commissioners and staff were involved in numerous public education activities domestically and abroad in which they discussed the commission's recommendations or work in progress. Those activities included meetings with congressional staff and federal agencies; participation in public lectures, seminars, and conferences; consultations with WHO, UNESCO, and CIOMS; and cosponsorship of the Second International Summit of National Bioethics Commissions in November 1998 in Tokyo. These activities have contributed to the policymaking process by increasing public awareness of bioethical issues.

Even though NBAC's tenure concluded, succeeding administrations and bioethics commissions, and in particular the President's Council on Bioethics, should be very interested in what NBAC recommended and the response to its work. As stated in NBAC's charter: "In order to avoid duplication of effort, the Commission is encouraged to review the deliberations of other entities." NBAC did just that during its deliberations. For example, NBAC considered the *Report of the Human Embryo Research Panel* (National Institutes of Health, 1994) when it addressed the issues of cloning human beings and human embryonic stem cell research. During its deliberations on *Research Involving Persons with Mental Disorders That May Affect Decision-making Capacity*, NBAC considered the National Commission's findings in the report *Research Involving Those Institutionalized as Mentally Infirm* (National Commission . . ., 1978) and ACHRE's *Final Report* (Advisory Committee on Human Radiation Experiments, 1995). In preparing the final *Ethical and Policy Issues in Research Involving Human Participants* report, NBAC looked to the work of several previous commissions, including the National Commission, the President's Commission, and ACHRE.

Cloning and human stem cell research continue to be hotly debated topics and are still being addressed by Congress, and the protection of human research participants is currently of interest to members of Congress and is being addressed by various groups. Therefore, NBAC's reports and recommendations continue to be relevant to current policy debates on bioethical issues.

Debate on many of the issues addressed by NBAC continues and will continue for some time to come. Therefore, this report constitutes an early assessment of the response to NBAC's work. Given the nature of the issues addressed by NBAC and the oftentimes slow pace of the policymaking process, the response to NBAC's reports and recommendations should be assessed over time to determine both their immediate and their long-term impact.

DATA COLLECTION METHODS

As part of an Intergovernmental Personnel Act agreement between RAND and NBAC, the author of this report tracked the response to NBAC's six reports, and the recommendations contained in those reports, from March 2000 until October 2001 when NBAC's charter expired. This study was conducted in an effort to assess NBAC's contribution to the policymaking process as it relates to various bioethical and scientific issues. This appendix describes the information sources that were searched and how the responses to the reports were tracked.

Chapter Two of this report describes the types of discussions about NBAC that appeared in the academic literature and in the media. The data in that chapter were obtained by searching the academic literature and the media for any mentions of NBAC, its reports, or its recommendations.

Data for the primary focus of this report—the response to NBAC's reports and recommendations, which are discussed in Chapters Three through Eight—were obtained by collecting documents that referred to, discussed, or were based on any of NBAC's reports and recommendations. Those documents included, but were not limited to, federal and state legislation; congressional testimony; presidential administration and federal agency guidelines, statements, policies, and procedures; statements from professional societies, organizations, and foundations; and laws, regulations, and guidelines from other countries and international organizations.

More-detailed information on the literature and media that were searched and the documents that were collected follows.

SEARCHING THE LITERATURE AND MEDIA FOR MENTIONS OF NBAC

Mentions of NBAC, its reports, or its recommendations in the academic literature or in the media were found through searches of relevant databases:

- Articles in the academic literature that mentioned NBAC and/or discussed its reports and recommendations were ascertained by searching the on-line databases PubMed, Bioethicsline, Social Science Index, and Science Citation Index for January 1996 through July 2001. Articles were also found through individual searches of the on-line versions of *Journal of the American Medical Association*, *New England Journal of Medicine*, *Science*, and *Nature* and other Nature Publishing Group journals. Short items, such as letters to the editor and articles appearing in news sections of journals, were not included.

- Media reports that mentioned NBAC and/or discussed its reports and recommendations were determined by a search of the Nexis news database from January 1, 1996, through June 30, 2001.

COLLECTING DOCUMENTS IN RESPONSE TO NBAC'S REPORTS AND RECOMMENDATIONS

Documents that referred to or were based on any of NBAC's reports and recommendations were obtained from several sources. The documents were collected from March 2000 until October 3, 2001, when NBAC's charter expired. The documents were obtained as follows:

- Administrative statements were sent to NBAC by the president of the United States and other individuals within the Clinton administration. RAND obtained these documents directly from NBAC.

- Federal legislation that mentioned NBAC was found through a search of the Library of Congress's THOMAS Legislative Information on the Internet Web site.[1] Individual searches of the leg-

---

[1]The THOMAS Legislative Information on the Internet Web site (http://thomas.loc. gov/) is a service the Library of Congress.

islation for the 104th, 105th, 106th, and 107th Congresses for the terms "National Bioethics Advisory Commission," "NBAC," and "bioethics" were performed. All relevant bills were identified.

- Federal department and agency policies, guidelines, statements, and other relevant materials that mentioned NBAC were found through searches of each agency's Web site for the terms "National Bioethics Advisory Commission" and "NBAC." Web sites for the following federal departments and agencies were searched: DHHS, CDC, FDA, NIH, Department of Defense, Department of Education, Department of Energy, Department of Veterans Affairs, and U.S. Agency for International Development. The specific departments and agencies were chosen primarily because of their involvement in research with human participants. Responses to NBAC's reports and recommendations were found for most of the departments and agencies whose Web sites were searched (all but the Department of Defense, Department of Education, and U.S. Agency for International Development).

- State legislation that mentions NBAC was found through a search of the Lexis database. Legislation introduced in Maryland and New York that was based on NBAC's recommendations in its *Research Involving Persons with Mental Disorders That May Affect Decisionmaking Capacity* report was obtained through Jack Schwartz, assistant attorney general and director of Health Policy Development, Maryland Attorney General's Office.

- Although it was not possible to identify every professional society, philanthropic foundation, organization representing private industry, and patient advocacy group that may have cited NBAC's reports and recommendations, several of those organizations were identified through their previous interactions with NBAC, such as testifying before NBAC and responding to public comment periods on NBAC's draft reports.

Web sites for the following professional and academic societies, associations, and organizations; philanthropic foundations and organizations; associations and organizations representing private industry; and patient advocacy groups were searched: Alzheimer's Association, American Association for the Advancement of Science, American College of Obstetricians and Gynecologists, American Society for Cell Biology, American Society for

Investigative Pathology, American Society for Reproductive Medicine, American Society of Human Genetics, Association of American Medical Colleges, Bill & Melinda Gates Foundation, Biotechnology Industry Organization, Elizabeth Glaser Pediatric AIDS Foundation, Endocrine Society, Federation of American Societies for Experimental Biology, Juvenile Diabetes Research Foundation International, The National Academies, National Organization for Rare Disorders, The Pew Charitable Trusts, Pharmaceutical Research and Manufacturers Association, Public Citizen, Society for Developmental Biology, and United States Conference of Catholic Bishops.

Policies, guidelines, statements, and other relevant materials that mention NBAC were found through searches of each organization's Web sites for the terms "National Bioethics Advisory Commission" and "NBAC." For organizations whose Web sites did not have search engines, relevant materials were identified by surfing the Web site and examining individual Web pages for references to NBAC. Responses to NBAC's reports and recommendations were found for 12 of the 21 organizations whose Web sites were searched.

- Policies, guidelines, statements, and other relevant materials from other countries that mentioned NBAC were found through Web searches for the terms "National Bioethics Advisory Commission" and "NBAC" both alone and in association with country names (e.g., "NBAC AND Canada" or "National Bioethics Advisory Commission AND Australia") using the Google.com search engine. The following countries were included in the search: Australia, Brazil, Canada, China, Denmark, Finland, France, India, Japan, the Netherlands, New Zealand, South Africa, Thailand, Uganda, and the United Kingdom. Responses to NBAC's reports and recommendations were found for about a third of these countries.

- Policies, guidelines, statements, and other relevant materials from international organizations that mention NBAC were found through searches of each organization's Web site for the terms "National Bioethics Advisory Commission" and "NBAC." For international organizations that did not have search engines on their Web sites, relevant materials were identified by surfing each Web site and examining individual Web pages for references to

NBAC. International organizations were selected primarily because of their contribution to policy on bioethical issues. Web
sites for the following international organizations were searched:
Council of Europe, CIOMS, EGE, European Commission, European Parliament, HUGO, International Conference on Harmo-
nization, OECD, UNAIDS, UNESCO, World Health Organization,
and World Medical Association. Responses to NBAC's reports
and recommendations were found for about a third of the international organizations that were searched.

BIBLIOGRAPHY

This Bibliography is divided into two sections. The first section lists the references cited in the chapters of this report, and the second is a compendium of the academic literature that has mentioned or discussed NBAC or its work (as discussed in Chapter Two of this report).

SOURCES CITED IN THIS REPORT

AAAS. *See* American Association for the Advancement of Science.

"AAMC Comment Letter on NIH Proposed Stem Cell Guidelines," letter to NIH Office of Science Policy, AAMC Government Affairs and Advocacy, January 19, 2000 (available at **http://www.aamc. org/advocacy/library/research/corres/2000/011900.h t m** as of March 2003).

"ACE, AAU, NASULGC Letter to HHS Secretary, with Transmittal Letters to the President and Congress, on NIH Guidelines for Embryonic Stem Cell Research," Association of American Universities, March 26, 2001 (available at **http://www.aau.edu/research/ StemCell3.26.01.html** as of March 2003).

Advisory Committee on Human Radiation Experiments, *Final Report,* Washington, D.C.: Government Printing Office, 1995.

Alliance for Human Research Protection, letter to the acting commissioner of the Food and Drug Administration on FDA Interim Rule, "Additional Safeguards for Children in Clinical Investigations of FDA-Regulated Products," August 6, 2001.

Alzheimer's Association, "Research Funded in 2000: Diagnosis, Treatment, and Prevention" Web site, n.d. (abstracts of funded research available at **http://www.alz.org/Researchers/Funded/ 2000/2000Diagnosis.html** as of March 2003).

American Association for the Advancement of Science and the Institute for Civil Society, *Stem Cell Research and Applications: Monitoring the Frontiers of Biomedical Research,* Washington, D.C.: AAAS and ICS, November 1999.

American Medical Association, Council on Scientific Affairs, *Cloning and Embryo Research,* Report 7 of the Council on Scientific Affairs (A-99), 1999a (available at **http://www.ama-assn.org/ama/pub/ article/2036-2503.html** as of March 2003).

American Medical Association, Council on Ethical and Judicial Affairs, *The Ethics of Human Cloning,* CEJA Report 2-A-99, 1999b, (available at **http://ama-assn.org/ama1/upload/mm/369/ report98.pdf** as of March 2003).

American Medical Association, Council on Scientific Affairs, *Embryonic/Pluripotent Stem Cell Research and Funding,* Report 15 of the AMA Council on Scientific Affairs (I-99), 1999c (available at **http:// ama-assn.org/ama/pub/print/article/2036-4106.html** as of March 2003).

American Medical Association, Council on Ethical and Judicial Affairs, *Ethical Considerations in International Research,* CEJA Report 117, 2001 (available at **http://www.ama-assn.org/ama/ upload/mm/369/report_117.pdf** as of March 2003).

American Society for Investigative Pathology, *National Bioethics Advisory Commission Report on "Cloning Human Beings,"* Bethesda, Md.: ASIP, 1997 (available at **http://asip.uthscsa.edu/PUBAFF_ ED/lynch.html** as of March 2003).

Angell, M., "The Ethics of Clinical Research in the Third World," *New England Journal of Medicine,* Vol. 337, No. 12, 1997, pp. 847–849.

"ASIP Seeks FASEB Support on Human Tissue Research Policy," *FASEB Newsletter,* Vol. 30, No. 1, February 1997 (available at **http://www.faseb.org/opar/newsletter/feb97/2_febx7.html** as of March 2003).

Association of American Medical Colleges, "Recent Activities Related to Human Subjects Protections," AAMC President's Memoranda Advisory, 1998.

"Assurance of Protection for Human Subjects," minutes of Human Subjects Research Advisory Committee, January 12, 2001 (available at **http://ohsr.od.nih.gov/HSRACminutes/011201_HSRAC_minutes.html** as of March 2003).

Australian Academy of Science, "On Human Cloning: A Position Statement," Canberra: Australian Academy of Science, February 1999.

Australian Health Ethics Committee, National Health and Medical Research Council, *Scientific, Ethical and Regulatory Considerations Relevant to Cloning Human Beings,* December 1998.

"Battle over Stem Cell Research Continues," *Science & Technology in Congress,* March 2000 (available at **http://www.aaas.org/spp/dspp/cstc/bulletin/articles/3-00/stem.htm** as of March 2003).

"BIO Responds to Cloning Report," press release, Biotechnology Industry Organization, Washington, D.C., January 7, 1998 (available at **http://www.bio.org/news/press_010798.html** as of March 2003).

"BIO Supports Ban on Cloning of Humans, Seeks Medical Research Protection," press release, Biotechnology Industry Organization, Washington, D.C., June 7, 1997 (available at **http://www.bio.org/news/june797.html** as of March 2003).

"BIO Urges President's Bioethics Advisory Panel To Consider Issues Raised By Stem Cell Research," press release, Biotechnology Industry Organization, Washington, D.C., November 12, 1998 (available at **http://www.bio.org/news/111298press.html** as of March 2003).

Biotechnology Industry Organization, "BIO's Recommendations for the National Bioethics Advisory Commission Regarding the Implications of Cloning Technology," Washington, D.C.: BIO, 1997 (available at **http://www.bio.org/bioethics/nbac.html** as of March 2003).

Biotechnology Industry Organization, "Cloning," Washington, D.C.: BIO, 1999a (available at **http://www.bio.org/bioethics/cloning_ paper1.html** as of March 2003).

Biotechnology Industry Organization, "Cloning," Washington, D.C.: BIO, 1999b (available at **http://www.bio.org/bioethics/cloning_ paper2.html** as of March 2003).

Biotechnology Industry Organization, *Editors' and Reporters' Guide to Biotechnology, Third Edition, 1998–1999,* Washington, D.C.: BIO, October 21, 1999c (available at **http://www.bio.org/ aboutbio/guide1.html** as of March 2003).

Biotechnology Industry Organization, *Editors' and Reporters' Guide to Biotechnology, Fourth Edition, 1999–2000,* Washington, D.C.: BIO, June 2000a (available at **http://www.bio.org/aboutbio/ guide1.html** as of March 2003).

Biotechnology Industry Organization, *Encouraging Development of the Biotechnology Industry: A Best Practices Survey of State Efforts, March 2000,* Washington, D.C.: BIO, April 11, 2000b.

Biotechnology Industry Organization, *Editors' and Reporters' Guide to Biotechnology, Fifth Edition, 2000–2001,* Washington, D.C.: BIO, June 2001 (available at **http://www.bio.org/aboutbio/guide1.html** as of March 2003).

Brady, R. P., M. S. Newberry, and V. W. Gerard, "FDA Regulatory Controls Over Human Stem Cells," *Professional Ethics Report,* Vol. 12, No. 2, Spring 1999 (available at: **http://aaas.org/spp/dspp/sfrl/ per/per17.htm** as of March 2003).

Byrnes, M. K., "Public Voices, Public Choices," The Pew Charitable Trusts Health and Human Services Program, Philadelphia, Pa.: Pew Charitable Trusts, June 1998.

Campbell, C. S., "Religious Perspectives on Human Cloning," in *Cloning Human Beings,* Vol. II, Rockville, Md.: National Bioethics Advisory Commission, June 1997.

Canadian Biotechnology Advisory Committee, *Annual Report 1999– 2000,* Ottawa, Ont.: CBAC, 2001 (available at **http://www.cbac-cccb.ca/epic/internet/incbac-cccb.nsf/vwapj/feb2001_Annual**

Report_e.pdf/$FILE/feb2001_AnnualReport_e.pdf as of March 2003).

Canadian Institutes of Health Research, "Working Paper: The Ethics Mandate of the Canadian Institutes of Health Research: Implementing a Transformative Vision," Ottawa, Ont., November 10, 1999.

Canadian Institutes of Health Research, "Human Stem Cell Research: Opportunities for Health and Ethical Perspectives: A Discussion Paper," Ottawa, Ont., March 29, 2001 (available at **http://www.cihr-irsc.gc.ca/publications/ethics/stem_cell/stem_cell_e.pdf** as of May 2003).

CDC. *See* Centers for Disease Control.

Center for Bioethics and Human Dignity, "On Human Embryos and Medical Research: An Appeal for Legally and Ethically Responsible Science and Public Policy," Bannockburn, Ill., July 1, 1999 (available at **http://www.cbhd.org/resources/aps/escs_99-07-01.htm** as of April 2003).

Centers for Disease Control and Prevention, "Translating Advances in Human Genetics into Public Health Action: A Strategic Plan," Atlanta, Ga.: CDC, October 1, 1997.

Centers for Disease Control and Prevention, National Institutes of Health, Food and Drug Administration, Health Resources and Services Administration, Substance Abuse and Mental Health Services Administration, and Indian Health Service, *Building Community Partnerships in Research*, Washington, D.C.: CDC, April 1998.

Centers for Disease Control and Prevention, "Notice of Intent; Genetic Testing Under the Clinical Laboratory Improvement Amendments," *Federal Register*, Vol. 65, No. 87, FR Doc. 00-11093, Thursday, May 4, 2000 (available at **http://www.lcc.gc.ca/en/themes/gr/hrish/macdonald/macdonald.pdf** as of April 2003).

Centre on Governance, University of Ottawa, *The Governance of Health Research Involving Human Subjects—Final Report*, Ottawa, Ont., March 15, 2000.

Chief Medical Officer, Department of Health, Department for Education and Employment Home Office, *The Removal, Retention and Use of Human Organs and Tissue from Post-mortem Examination: Advice from the Chief Medical Officer,"* London, UK: The Stationery Office, 2001 (available at **http://www.doh.gov.uk/cegc/stemcellreport.pdf** as of April 2003).

Chief Medical Officer's Expert Group Reviewing the Potential of Developments in Stem Cell Research and Cell Nuclear Replacement to Benefit Human Health, Department of Health, *Stem Cell Research: Medical Progress with Responsibility,* London, UK: The Stationary Office, June 2000.

Chow, I., "Update on Human Pluripotent Embryonic Stem Cell Research: What's Happening?" Society for Developmental Biology, February 1999 (available at **http://sdb.bio.purdue.edu/publications/focus/stem_cell_article.html** as of March 2003).

"Cloning: From DNA Molecules to Dolly," *Human Genome News,* Vol. 9, Nos. 1–2, January 1998 (available at **http://www.ornl.gov/TechResources/Human_Genome/publicat/hgn/v9n1/01tocont.html** as of March 2003).

"Cloning Prohibition Act of 1997," message from the President of the United States to the U.S. Senate, PM 46, June 9, 1997a.

"Cloning Prohibition Act of 1997," message from the President of the United States to the U.S. House of Representatives, H. DOC. No. 105-97, June 10, 1997b.

Coleman, C., "Drug Research for Mental Illness: The Debate Is Highly Charged, But What's Really at Stake?" *AAMC Reporter,* Vol. 8, No. 8, 1999.

"Comments to President Clinton Regarding NBAC," letter from the Presidents of the American Academy of Neurology and the American Neurological Association, April 14, 1999.

"The Complete Guide to Cloning," *Life Insight,* Vol. 12, No. 2, March–April 2001 (available at **http://www.usccb.org/prolife/publicat/lifeinsight/marapr2001.htm** as of March 2003).

Council of Europe, "The Protection of Medical Data, Recommendation No. R (97)5 and explanatory memorandum," 1997 (available at **http://book.coe.int/gb/cat/liv/htm/11097.htm** as of March 31, 2003).

Council for Science and Technology, Human Embryo Research Subcommittee of the Bioethics Committee, *Basic Viewpoints on Human Embryo Research Which Centers on Research Involving Human Embryonic Stem Cells,* London, UK: CST, March 2000.

"Debate over Embryonic Stem Cells Grows," *Science & Technology in Congress,* July 1999 (available at **http://www.aaas.org/spp/dspp/ cstc/bulletin/articles/7-99/stcells.htm** as of March 2003).

Di Berardino, M. A., "Cloning: Past, Present, and the Exciting Future," *Breakthroughs in Bioscience,* Bethesda, Md.: Federation of American Societies for Experimental Biology (available at **http://www.faseb.org/opar/cloning and http://www.faseb.org/ opar/cloning/cloning/pdf** as of March 2003).

Division of Microbiology and Infectious Diseases, National Institute of Allergy and Infectious Diseases, "Suggested Language for Informed Consent for Future Use of Biological Specimens Collected Under Clinical Protocols," in "Guidance for Clinical Protocol Development," in "Request for Proposal: Hepatitis C Recovery Research Network (HC RRN)," request for proposal NIH-NIAID-DNID-01-16, July 26, 2000. (available at **http://www.niaid.nih. gov/contract/archive/ACF275.pdf** as of March 2003).

Eiseman, E., and S. B. Haga, *Handbook of Human Tissue Sources: A National Resource of Human Tissue Samples,* Santa Monica, Calif.: RAND, MR-954, 1999.

"Embryonic Stem Cell Research: A History," *Science & Technology in Congress,* September 2001 (available at **http://www.aaas.org/spp/ cstc/stc/stc01/01-09/stem2.htm** as of March 2003).

Endocrine Society, *Code of Ethics of the Endocrine Society,* 2001 (available at **http://www.endo-society.org/pubrelations/ ee20018398.pdf** as of March 2003).

English, V., and V. Nathanson, "Human 'Cloning'—A Discussion Paper for the World Medical Association," British Medical Association, October 1999.

European Group on Ethics in Science and New Technologies to the European Commission, "Opinion No. 15: Ethical Aspects of Human Stem Cell Research and Use," *General Report on the Activities of the European Group on Ethics in Science and New Technologies to the European Commission 1998–2000*, 2000, pp. 121–135 (available at **http://europa.eu.int/comm/european_group_ethics/docs/avis15_en.pdf** as of March 31, 2003).

European Parliament Directorate General for Research, *The Ethical Implications of Research Involving Human Embryos: Final Study*, Luxembourg, Belgium, July 2000.

"FASEB Endorses Voluntary Moratorium on Cloning Human Beings," *FASEBnews*, September 18, 1997 (available at **http://www.faseb.org/opar/cloning.moratorium.html** as of March 2003).

Feldbaum, Carl B., BIO president, letter to President George W. Bush, February 1, 2001, Washington, D.C.: Biotechnology Industry Organization (available at **http://www.bio.org/bioethics/cloning_letter_bush.html** as of March 2003).

Foubister, V., "Patients Should Be the Ones to Profit from Trials," *American Medical News* and *amednews.com*, July 10/17, 2000 (available at **http://www.ama-assn.org/sci-pubs/amnews/pick_00/prsb0710.htm** as of March 2003).

Foubister, V., "National Bioethics Advisory Commission Sets Protections for Clinical Trial Subjects," *American Medical News* and *amednews.com*, February 19, 2001 (available at **http://www.ama-assn.org/sci-pubs/amnews/pick_01/prsb0219.htm** as of March 2003).

Garfinkel, M., "Stem Cell Controversy," *Professional Ethics Report* XII(3), Summer 1999 (available at **http://aaas.org/spp/dspp/sfrl/per/per18.htm** as of March 2003).

Gianelli, D. M., "Panel Takes Middle Road on Stem Cell Research," *American Medical News* and *amednews.com*, August 2, 1999 (available at **http://www.ama-assn.org/sci-pubs/amnews/ pick_99/prfa0802.htm** as of March 2003).

HHS Working Group on the NBAC Report, U.S. Department of Health and Human Services, "Analysis and Proposed Actions Regarding the NBAC Report: Research Involving Persons with Mental Disorders That May Affect Decisionmaking Capacity," Washington, D.C.: DHHS, January 16, 2001.

Hübner, D., Institut für Wissenschaft und Ethik, correspondence with Eric Meslin, executive director of NBAC, October 13, 2000.

HUGO Ethics Committee, "Statement on the Principled Conduct of Genetic Research," March 1996 (available at **http://www.hugo-international.org/hugo/conduct.htm** as of March 31, 2003).

Human Genetics Advisory Commission and Human Fertilisation and Embryology Authority, "Cloning Issues in Reproduction, Science and Medicine," HGAC paper, December 1988 (available at **http://www.doh.gov.uk/hgac/papers/paperd1.htm** as of March 2003).

"Human Trials Protected in International Research and Drug Trials," *Professional Ethics Report,* Vol. XIII, No. 4, Fall 2000 (available at **http://www.aaas.org/spp/dspp/sfrl/per/per23.htm** as of April 2003).

Hyman, S., "Importance of Clinical Research and Protecting Human Subjects," National Institute of Mental Health 3rd Annual Research Roundtable, June 23, 1999 (available at **http://www.nimh. nih.gov/events/roundtable.htm** as of March 2003).

Institut für Wissenschaft und Ethik, *Jahrbuch für Wissenschaft und Ethik* (Yearbook for Science and Ethics), Bonn, Germany: Institut für Wissenschaft und Ethik, 2000.

International Bioethics Committee, United Nations Educational, Scientific and Cultural Organization, *The Use of Embryonic Stem Cells in Therapeutic Research: Report of the IBC on the Ethical Aspects of Human Embryonic Stem Cell Research,* Paris: IBC, April 6, 2001 (available at **http://www.unesco.org/ibc/en/reports/embryonic_ ibc_report.pdf** as of March 2003).

Jaenisch, R., American Society for Cell Biology, Bethesda, Md., testimony to the Science, Technology and Space Subcommittee of the Senate Commerce, Science and Transportation Committee, U.S. Senate, May 2, 2001 (available at **http://www.ascb.org/ publicpolicy/clonetest.htm** as of March 2003).

John E. Fogarty International Center, National Institutes of Health, "Minutes of the Advisory Board, Forty-Fifth Meeting," May 16, 2000 (available at **http://www.fic.nih.gov/about/ minadv20000516.html** as of April 2003).

Kirschtein, R., acting director, National Institutes of Health, "NIH and the Academic Research Community: Partnerships for Clinical Trials," Association of American Medical Colleges Grand Annual Meeting, Washington, D.C., May 7, 2000 (available at **http://www. nih.gov/about/director/Speeches/aamc57.htm** as of March 2003).

Koizumi, K., and P. W. Turner, "Agency R&D Budgets," in *Congressional Action on Research & Development in the FY 2000 Budget,* Washington, D.C.: American Association for the Advancement of Science, 2000.

Lanzendorf, S. E., C. A. Boyd, D. L. Wright, S. Muasher, S. Oehninger, and G. D. Hodgen, "Use of Human Gametes Obtained from Anonymous Donors for the Production of Human Embryonic Stem Cell Lines," *Fertility and Sterility*, Vol. 76, No. 1, 2001, pp. 132–137.

Lurie, P., and S. Wolfe, "Unethical Trials of Interventions to Reduce Perinatal Transmission of the Human Immunodeficiency Virus in Developing Countries," *New England Journal of Medicine*, Vol. 337, No. 12, 1997, pp. 853–856.

Mann, H., "Research Using Human Biological Materials: IRB Policies and Procedures," University of Utah School of Medicine and LDS Hospital, Intermountain Health Care, Salt Lake City, n.d. (available at **http://www.ihc.com/xp/ihc/lds/forresearchers/irb/ humanbiomatresearch/** as of March 2003).

Murray, T. H., *Ethical Challenges in Research with Human Biological Materials,* online teaching module, Online Ethics Center for

Engineering and Science at Case Western Reserve University, n.d. (available at **http://www.onlineethics.org/reseth/mod/biores. html** as of March 2003).

National Academies, *Stem Cells and the Future of Regenerative Medicine,* Washington, D.C.: National Academy Press, 2001.

National Bioethics Advisory Commission, *Cloning Human Beings,* Rockville, Md.: NBAC, June 1997 (available at **http://www. georgetown.edu/research/nrcbl/nbac/pubs.html** as of March 2003).

National Bioethics Advisory Commission, *Research Involving Persons with Mental Disorders That May Affect Decisionmaking Capacity,* Rockville, Md.: NBAC, December 1998 (available at **http://www. georgetown.edu/research/nrcbl/nbac/pubs.html** as of March 2003).

National Bioethics Advisory Commission, *Research Involving Human Biological Materials: Ethical Issues and Policy Guidance,* Rockville, Md.: NBAC, August 1999a (available at **http://www.georgetown. edu/research/nrcbl/nbac/pubs.html** as of March 2003).

National Bioethics Advisory Commission, *Ethical Issues in Human Stem Cell Research,* Rockville, Md.: NBAC, September 1999b (available at **http://www.georgetown.edu/research/nrcbl/nbac/ pubs.html** as of March 2003).

National Bioethics Advisory Commission, *Ethical and Policy Issues in International Research: Clinical Trials in Developing Countries,* Rockville, Md.: NBAC, April 2001a (available at **http://www. georgetown.edu/research/nrcbl/nbac/pubs.html** as of March 2003).

National Bioethics Advisory Commission, *Ethical and Policy Issues in Research Involving Human Participants,* Rockville, Md.: NBAC, August 2001b (available at **http://www.georgetown.edu/research/ nrcbl/nbac/pubs.html** as of March 2003).

"National Bioethics Advisory Commission (NBAC) Scrutinizes Current System for Protecting Research Subjects," *news@vhaethics,* National Center for Ethics, Winter 2001.

"National Call to Ban Human Cloning," *Professional Ethics Report*, Vol. 10, No. 2, American Association for the Advancement of Science, Spring 1997 (available at **http://aaas.org/spp/dspp/sfrl/per/ per9.htm** as of March 2003).

National Cancer Institute, Division of Extramural Activities, "Board of Scientific Advisors Meeting Minutes," Bethesda, Md., March 2–3, 1998, (available at **http://deainfo.nci.nih.gov/advisory/bsa/ bsa0398/bsa0398.htm** as of March 2003).

National Commission for the Protection of Human Subjects of Biomedical and Behavioral Research (National Commission), *Research Involving Those Institutionalized as Mentally Infirm*, Washington, D.C.: Department of Health, Education and Welfare, 1978.

National Commission for the Protection of Human Subjects of Biomedical and Behavioral Research (National Commission), *Belmont Report: Ethical Principals and Guidelines for the Protection of Human Subjects of Research*, Washington, D.C.: U.S. Government Printing Office, 1979.

National Heart, Lung, and Blood Institute, National Institutes of Health, *Human Tissue Repository Guidelines*, April 14, 2000 (available at **http://www.nhlbi.nih.gov/funding/policies/repos-gl.htm** as of March 2003).

National Human Genome Research Institute, Ethical, Legal, and Social Implications Programs, Research Planning and Evaluation Group, *A Review and Analysis of the ELSI Research Programs at the National Institutes of Health and the Department of Energy: Final Report of the ELSI Research Planning and Evaluation Group*, February 10, 2000 (available at **http://www.genome.gov/10001727** as of April 2003).

National Institute of Diabetes and Digestive and Kidney Diseases, Stem Cell and Developmental Biology Planning Group, *Stem Cell and Developmental Biology Writing Group's Report—Draft Version 1.0*, National Institutes of Health, Bethesda, Md., n.d. (available at **http://www.niddk.nih.gov/federal/planning/ stemcell-writing.htm** as of March 2003).

National Institute on Drug Abuse, "NACDA Guidelines for the Administration of Drugs to Human Subjects," approved by the National Advisory Council on Drug Abuse on September 13, 2000 (available at **http://www.nida.nih.gov/Funding/HSGuide.html** as of March 2003).

National Institute of General Medical Sciences, *Population-Based Samples for the NIGMS Human Genetic Cell Repository 7-20-99*, Bethesda, Md.: National Institute of General Medical Sciences, July 20, 1999 (available at **http://www.nigms.nih.gov/news/reports/cellrepos.html** March 2003).

National Institute of General Medical Sciences, *Report of the First Community Consultation on the Responsible Collection and Use of Samples for Genetic Research*, Bethesda, Md.: National Institute of General Medical Sciences, September 25–26, 2000 (available at **http://www.nigms.nih.gov/news/reports/community_consultation.html** as of March 2003).

National Institute of Mental Health, National Institutes of Health, *Director's Report to the National Advisory Mental Health Council*, September 18, 1998 (available at **http://www.nimh.nih.gov/council/dir998.cfm** as of March 2003).

National Institute of Mental Health, "Minutes of the 191st Meeting of the National Advisory Mental Health Council," February 4–5, 1999 (available at **http://www.nimh.nih.gov/council/min299.cfm** as of March 2003).

National Institute of Mental Health, *Issues to Consider in Intervention Research with Persons at High Risk for Suicidality*, Bethesda, Md.: NIMH, January 2001 (available at **http://www.nimh.nih.gov/research/highrisksuicide.cfm** as of March 2003).

National Institute of Neurological Disorders and Stroke, "NINDS Advisory Council Meeting Minutes," meetings of the National Advisory Neurological Disorders and Stroke Council, February 11–12, 1999 (available at **http://www.ninds.nih.gov/about_ninds/council_minutes_february1999.htm** as of March 2003), and May 20–21, 1999 (available at **http://www.ninds.nih.gov/about_ninds/council_minutes_may1999.htm** as of March 2003).

National Institutes of Health, *Report of the Human Embryo Research Panel,* Bethesda, Md.: NIH, 1994.

National Institutes of Health, *Draft National Institutes of Health Guidelines for Stem Cell Research,* in *Federal Register,* Vol. 64, No. 231, December 2, 1999, pp. 67576–67579 (available at **http://www.nih.gov/news/stemcell/draftguidelines.htm** as of March 2003).

National Institutes of Health, *National Institutes of Health Guidelines for Research Using Human Pluripotent Stem Cells,* in *Federal Register,* Vol. 65, No. 166, August 25, 2000, pp. 51975–51981; corrected in *Federal Register,* Vol. 65, No. 225, November 21, 2000, p. 69951 (available at **http://www.nih.gov/news/stemcell/stemcellguidelines.htm** as of March 2003).

National Institutes of Health, "NIH Fact Sheet on Human Pluripotent Stem Cell Research Guidelines," Bethesda, Md.: NIH, updated January 2001 (available at **http://www.nih.gov/news/stemcell/stemfactsheet.htm** as of March 2003).

"NBAC Considers Expansion of IRB Regulations," *AAMC Washington Highlights,* April 14, 2000.

"NBAC Continues Discussions on Human Subjects Research," *AAMC Washington Highlights,* July 21, 2000.

"NBAC Draft Oversight Report/Memo to Partners," CDC Prevention News Update, American Medical Association, February 7, 2001 (available at **http://www.ama-assn.org/ama/pub/article/3608-3858.html** as of April 2003).

"NBAC Issues Report on International Research," *Professional Ethics Report,* Vol. XIV, No. 2, Spring 2001 (available at **http://www.aaas.org/spp/dspp/sfrl/per/per25.htm** as of April 2003).

"NBAC Proposes Human Subject Reforms," *Science & Technology in Congress,* Center for Science, Technology, and Congress, June 2001.

Nelson, S. D., and D. G. Cooper, "Political and Policy Context for the FY 2002 Budget," in *AAAS Report XXVI: Research and Development*

FY 2002, Intersociety Working Group, Washington, D.C.: American Association for the Advancement of Science, 2001.

The NIAID Division of AIDS, "Ethical Issues in HIV Prevention Trials," *Science* Web site, n.d. (available at **http://www.niaid.nih.gov/daids/prevention/ethics.htm** as of April 2003).

Nightingale, S. L., FDA associate commissioner, "Dear Colleague" letter, October 26, 1998 (available at **http://www.fda.gov/oc/ohrt/irbs/irbletr.html** as of March 2003).

NIH Director's Council of Public Representatives, "Spring 2000 Meeting Minutes," Bethesda, Md., April 6–7, 2000a (available at **http://www.nih.gov/about/publicliaison/040600minutes.htm** as of March 2003).

NIH Director's Council of Public Representatives, "Fall 2000 Meeting Minutes," Bethesda, Md., October 31–November 1, 2000b (available at **http://www.nih.gov/about/publicliaison/103100minutes.htm** as of March 2003).

NIH Director's Council of Public Representatives, "Spring 2001 Meeting Minutes," Bethesda, Md., May 1, 2001 (available at **http://www.nih.gov/about/publicliaison/COPRMinutesMay2001.htm** as of March 2003).

"NIH Publishes Draft Guidelines for Stem Cell Research," National Institutes of Health, NIH press release, December 1, 1999 (available at **http://www.nih.gov/news/pr/dec99/od-01.htm** as of March 2003).

"NIH Publishes Final Guidelines for Stem Cell Research," NIH press release, August 23, 2000 (available at **http://www.nih.gov/news/pr/aug2000/od-23.htm** as of March 2003).

"NIH Releases Guidelines for Stem Cell Research," *Professional Ethics Report*, Vol. XII, No. 4, American Association for the Advancement of Science, Fall 1999 (available at: **http://aaas.org/spp/dspp/sfrl/per/per19.htm** as of March 2003).

"Not So Fast, Dolly," *American Medical Association News* and *amednews.com*, September 27, 1999 (available at **http://www.ama-**

assn.org/sci-pubs/amnews/amn_99/edit0927.htm as of March 2003).

Nuffield Council on Bioethics, minutes of meetings of the Working Party on the Ethics of Healthcare-Related Research in Developing Countries meetings, December 7, 2000a (available at **http://www. nuffieldbioethics.org/developingcountries/mn_0000000068.asp** as of April 2003), and February 2, 2001 (available at **http://www. nuffieldbioethics.org/developingcountries/mn_0000000067.asp** as of April 2003).

Nuffield Council on Bioethics, "Stem Cell Therapy: The Ethical Issues," London, UK: Nuffield Council on Bioethics, April 6, 2000b (available at **http://www.nuffieldbioethics.org/publications/ pp_0000000007.asp** as of March 2003).

Office of the Director, National Institutes of Health, "Proceedings of the 78th Meeting of the Advisory Committee to the Director," report from the Stem Cell Working Group, June 3, 1999a (available at **http://www.nih.gov/about/director/minutes699.htm** as of March 2003).

Office of the Director, National Institutes of Health, "Proceedings of the 79th Meeting of the Advisory Committee to the Director," update on Stem Cell Research, December 2, 1999b (available at **http://www.nih.gov/about/director/dec99min.htm** as of March 2003).

Office of Extramural Research, National Institutes of Health, "Research Involving Individuals with Questionable Capacity to Consent: Points to Consider," March 11, 1999 (available at **http://grants.nih.gov/grants/policy/questionablecapacity.htm** as of March 2003).

Office of Human Subjects Research, National Institutes of Health, *Guidance on the Research Use of Stored Samples or Data,* Information Sheet 14, August 2000 (available at **http://ohsr.od.nih. gov/info/ninfo_14.php3** as of March 2003).

Office of the Inspector General, Department of Health and Human Services, *Institutional Review Boards: A Time for Reform,* OEI-01-

97-00193, June 1998 (available at **http://oig.hhs.gov/oei/ reports/oei-01-97-00193.pdf** as of March 2003).

Office of the Inspector General, Department of Health and Human Services, *Protecting Human Research Subjects: Status of Recommendation*, OEI-01-97-00197, April 2000 (available at **http:// oig.hhs.gov/oei/reports/oei-01-97-00197.pdf** as of March 2003).

Office for Protection from Research Risks, *Report to the Advisory Committee to the Director, NIH, from the Office for Protection from Research Risks Review Panel*, June 3, 1999 (available at **http:// www.nih.gov/about/director/060399b.htm** as of March 2003).

Okarma, T., president and CEO of Geron Corporation, testimony on behalf of BIO before the Subcommittee on Oversight and Investigations, Committee on Energy and Commerce, U.S. House of Representatives, March 28, 2001 (available at **http://www. bio.org/bioethics/okarmatestimony.htm** as of March 2003).

Okazaki, Y., director of the R&D Committee of the Japan Pharmaceutical Manufacturers Association, correspondence with Eric Meslin, executive director of NBAC, January 21, 2000.

Organization for Economic Co-operation and Development, *Genetic Testing: Policy Issues for the New Millennium*, Paris, France: OECD, 2000.

Parliament of Australia, Commonwealth of Australia, *The House of Representatives Legal and Constitutional Affairs Committee Bulletin*, n.d.

Pew Forum on Religion and Public Life, "Ethics of Human Cloning," Washington, D.C.: Pew Forum, May 2001.

"President's Bill Would Prohibit Human Cloning," *Human Genome News*, Vol. 8, Nos. 3–4, January–June 1997 (available at **http://www.ornl.gov/TechResources/Human_Genome/publicat/ hgn/v8n3/01tocont.html** as of March 2003).

"President's Commission Issues Cloning Recommendations," *Science & Technology in Congress*, AAAS, July 1997 (available at **http://www.aaas.org/spp/dspp/cstc/bulletin/articles/7-97/ CLONING.HTM** as of March 2003).

President's Commission for the Study of Ethical Problems in Medicine and Biomedical and Behavioral Research (President's Commission), *Protecting Human Subjects: First Biennial Report on the Adequacy and Uniformity of Federal Rules and Policies, and of Their Implementation, for the Protection of Human Subjects,* Washington, D.C.: U.S. Government Printing Office, 1981.

President's Commission for the Study of Ethical Problems in Medicine and Biomedical and Behavioral Research (President's Commission), *Implementing Human Research Regulations: Second Biennial Report on the Adequacy and Uniformity of Federal Rules and Policies, and of Their Implementation, for the Protection of Human Subjects,* Washington, D.C.: U.S. Government Printing Office, 1983.

"Protecting Workers with Developmental Disabilities," *Applied Occupational and Environmental Hygiene,* National Institute for Occupational Safety and Health, Vol. 15, No. 2, June 2000, pp. 171–181.

Public Citizen, letter to President William J. Clinton, June 11, 1997a (available at **http://www.publiccitizen.org/publications/release. cfm?ID=6616** as of April 2003).

Public Citizen, letter to Secretary Donna Shalala, April 22, 1997b (available at **http://www.publiccitizen.org/publications/release. cfm?ID=4896** as of April 2003).

Public Citizen, The Health Research Group, "Letter to the National Bioethics Advisory Commission Criticizing Their Draft Report on Ethics of Research in Developing Countries," letter to Harold T. Shapiro, chairman of NBAC, December 6, 2000 (available at **http://www.publiccitizen.org/publications/release.cfm?ID=6751** as of April 2003).

Public Citizen, letter to Tommy Thompson, Secretary, Department of Health and Human Services, February 22, 2001a (available at **http://www.publiccitizen.org/publications/release.cfm?ID=6761** as of April 2003).

Public Citizen, "National Bioethics Advisory Commission Report Dangerously Weakens International Protections: Report Provides

Less Protection for Participants in International Research Than Declaration of Helsinki," statement issued April 30, 2001b (available at **http://www.publiccitizen.org/pressroom/release. cfm?ID=619** as of April 2003).

Public Health Genetics Unit, "Stem Cells and Cloning: Brief UK Policy Update," Cambridge, UK: PHGU, n.d. (available at **http://www. medinfo.cam.ac.uk/phgu/info_database/ELSI/stem_cells.asp** as of March 2003).

Public Health Service, "Clinical Investigations Using Human Subjects," in *Final Report, Supplemental Vol. I,* Washington, D.C.: U.S. Government Printing Office, 1966.

"Remarks by the President in Apology for Study Done in Tuskegee," The White House, Office of the Press Secretary, press release, May 16, 1997 (available at **http://clinton4.nara.gov/textonly/New/ Remarks/Fri/19970516-898.html** as of April 2003).

Roehr, B., "Doing the Right Thing—Globally," *Journal of the International Association of Physicians in AIDS Care,* April 2000.

Secretary's Advisory Committee on Genetic Testing, *Enhancing the Oversight of Genetic Tests: Recommendations of the SACGT,* Bethesda, Md.: SACGT, June 2000.

Shamblott, M. J., J. Axelman, S. Wang, E. M. Bugg, J. W. Littlefield, P. J. Donovan, P. D. Blumenthal, G. R. Higgins, and J. D. Gearhart, "Derivation of Pluripotent Stem Cells from Cultured Human Primordial Germ Cells," *Proceedings of the National Academy of Sciences of the United States of America,* Vol. 95, 1998, pp. 13726– 13731.

Shore, D., and S. E. Hyman, "An NIMH Commentary on the NBAC Report," *Biological Psychiatry,* Vol. 46, 1999, pp. 1013–1016.

Society for Developmental Biology, "Human Embryonic Stem Cell Research—Where Does It Stand Now?" August 1999 (available at **http://sdb.bio.purdue.edu/publications/focus/stem_cell_899. html** as of March 2003).

Sotis, J. J., "Reflections on the United States National Institutes for Health, Draft Guidelines for Research Involving Human Pluripotent Stem Cells," *Medical Ethics and Bioethics* (*Medicínska Etika & Bioetika*), Bratislava, Slovak Republic, Vol. 7, Nos. 1–2, 2000, pp. 19–21.

"Standards for International Research Still a Work in Progress," *American Medical News*, February 12, 2001 (available at **http://www.ama-assn.org/sci-pubs/amnews/pick_01/prsb0212. htm** as of March 2003).

"Summary Minutes of the Microbiology Devices Panel Meeting," Open Session, Gaithersburg, Md., July 27–28, 2000 (available at **http://www.fda.gov/ohrms/dockets/ac/00/minutes/3637m1.pdf** as of March 2003).

Termeer, H. A., BIO chairman, and C. B. Feldbaum, BIO president, letter to President William J., Clinton, from the Biotechnology Industry Organization, Washington, D.C., March 27, 1997 (available at **http://www.bio.org/bioethics/cloning_letter_ clinton.html** as of March 2003).

Thomson, J. A., J. Itskovitz-Eldor, S. S. Shapiro, M. A. Waknitz, J. J. Swiergiel, V. S. Marshall, and J. M. Jones, "Embryonic Stem Cell Lines Derived from Human Blastocysts," *Science*, Vol. 282, 1998, pp. 1145–1147.

UNESCO. *See* United Nations Economic, Scientific and Cultural Organization.

United Nations Economic, Scientific and Cultural Organization, *Records of the 29th Session of the General Conference: Reports of the Programme Commissions, the Administrative Commission, the Joint Meeting of the Programme and Administrative Commissions, and the Legal Committee,* Vol. 2, Paris, France: UNESCO, 1997.

United Nations Educational, Scientific. and Cultural Organization, "Reproductive Human Cloning: Ethical Questions," June 1998.

United States v. Karl Brandt et al. ("The Medical Case"), *Trials of War Criminals Before the Nuremberg Military Tribunals Under Control Council Law No. 10,* Vols. I–II, Nuremberg, October 1946–April 1949, Washington, D.C.: U.S. Government Printing Office, 1949.

U.S. Food and Drug Administration, "Use of Cloning Technology to Clone a Human Being," CBER Web page, 2002 (available at **http://www.fda.gov/cber/genetherapy/clone.htm** as of March 2003).

Wade, N., "Human Cells Revert to Embryo State, Scientists Assert," *The New York Times*, November 12, 1998.

Wade, N., "U.S. to Pay for Embryo Stem Cell Research: New Rules Set Limits for Use of Tissue Taken from Fertility Clinics," *New York Times News Service*, August 24, 2000.

Weiss, R., "Scientists Use Embryos Made Only for Research," *The Washington Post*, July 11, 2001a, p. A01.

Weiss, R., "Firm Aims to Clone Embryos for Stem Cells," *The Washington Post*, July 12, 2001b, p. A01.

White House, Office of the Press Secretary, "Memorandum for the Heads of Executive Departments and Agencies," Washington, D.C., March 4, 1997.

Zoon, K. C., director of CBER, statement before the Subcommittee on Oversight and Investigations of the House Committee on Energy and Commerce, March 28, 2001a (available at **http://www.fda.gov/ola/2001/humancloning.html** as of March 2003).

Zoon, K. C., director of CBER, presentation to the FDA Science Board, April 13, 2001b (available at **http://www.fda.gov/ohrms/dockets/ac/01/transcripts/3741t1.rtf** as of March 2003).

ACADEMIC LITERATURE CITING NBAC

Annas, G. J., "Why We Should Ban Human Cloning," *New England Journal of Medicine*, 339:122–125, 1998.

Annas, G. J., "Ulysses and the Fate of Frozen Embryos—Reproduction, Research, or Destruction?" *New England Journal of Medicine*, 343:373–376, 2000.

Annas, G. J., A. Caplan, and S. Elias, "Stem Cell Politics, Ethics and Medical Progress," *Nature Medicine*, 5(12):1339–1341, 1999.

Appelbaum, P. S., "Competence and Consent to Research: A Critique of the Recommendations of the National Bioethics Advisory Commission," *Accountability in Research,* 7(2–4):265–276, 1999.

Ashburn, T. T., S. K. Wilson, and B. I. Eisenstein, "Human Tissue Research in the Genomic Era of Medicine: Balancing Individual and Societal Interests," *Archives of Internal Medicine,* 160(22):3377–3384, 2000.

Baylis, F., "Our Cells/Ourselves: Creating Human Embryos for Stem Cell Research," *Women's Health Issues,* 10(3):140–145, 2000.

Bereano, P. L., "The National Bioethics Advisory Commission Report on the Use of Human Biological Materials and Research: Ethical Issues and Policy Guidelines," *Biotechnology Law Report,* 18(4):322–325, 1999.

Berg, P., and M. Singer, "Regulating Human Cloning," *Science,* 282: 413, 1998.

Blacksher, E., "Cloning Human Beings: Responding to the National Bioethics Advisory Commission's Report," *Hastings Center Report,* 27(5):6–9, 1997.

Bloom, F. E., "Breakthroughs 1997," *Science,* 278: 2029, 1997.

Burris, J., R. Cook-Deegan, and B. Alberts, "The Human Genome Project After a Decade: Policy Issues," *Nature Genetics,* 20:333–335, 1998.

Callahan, D., "Cloning: The Work Not Done," *Hasting Center Report,* 27(5):18–20, 1997.

Campbell, C. S., "Religion and the Body in Medical Research," *Kennedy Institute of Ethics Journal,* 8(3):275–305, 1998.

Campbell, C. S., "In Whose Image: Religion and the Controversy of Human Cloning," *Second Opinion,* 1:24–43, 1999.

Capron, A. M., "An Egg Takes Flight: The Once and Future Life of the National Bioethics Advisory Commission, *Kennedy Institute of Ethics Journal,* 7(1):63–80, 1997.

Capron, A. M., "Inside the Beltway Again: A Sheep of a Different Feather," *Kennedy Institute of Ethics Journal*, 7(2):171–179, 1997.

Capron, A. M., "Ethical and Human-Rights Issues in Research on Mental Disorders That May Affect Decision-Making Capacity," *New England Journal of Medicine*, 340:1430–1434, 1999.

Caulfield, T., "Cloning and Genetic Determinism—A Call for Consistency," *Nature Biotechnology*, 19(5):403, 2001.

Charney, D. S., "The National Bioethics Advisory Commission Report: The Response of the Psychiatric Research Community Is Critical to Restoring Public Trust," *Archives of General Psychiatry*, 56(8):699–700, 1999.

Charo, R. A., "Update on the National Bioethics Advisory Commission," *Protecting Human Subjects*, Fall 1997.

Childress, J. F., "The Challenges of Public Ethics: Reflections on NBAC's Report," *Hasting Center Report*, 27(5):9–11, 1997.

Childress, J. F., "An Introduction to NBAC's Report on Research Involving Persons with Mental Disorders That May Affect Decision-making Capacity," *Accountability in Research*, 7(2–4):101–115, 1999.

Childress, J. F., and H. T. Shapiro, "Almost Persuaded: Reactions to Oldham et al.," *Archives of General Psychiatry*, 56(8):697–698, 1999.

Cohen, C. B., "Unmanaged Care: The Need to Regulate New Reproductive Technologies in the United States," *Bioethics*, 11(3–4): 348–365, 1997.

Cole-Turner, R., *Human Cloning: Religious Responses*, Louisville, Ky.: Westminster John Knox Press, 1997.

Collmann, J., and G. Graber, "Developing a Code of Ethics for Human Cloning," *Critical Reviews in Biomedical Engineering*, 28(3–4):563–566, 2000.

Emanuel, E. J., D. Wendler, and C. Grady, "What Makes Clinical Research Ethical?" *Journal of the American Medical Association*, 283(20):2701–2711, 2000.

"Executive Summary of Cloning Human Beings: Report and Recommendations of the National Bioethics Advisory Commission," *Issues in Law & Medicine,* 14(2):217–222, 1998.

Faden, R., "The Advisory Committee on Human Radiation Experiments: Reflections on a Presidential Commission," *Hastings Center Report,* 26(5):5–10, 1996.

Frankel, M. S., "In Search of Stem Cell Policy," *Science,* 287:1397, 2000.

Gordon, J. W., "Genetic Enhancement in Humans," *Science,* 283: 2023–2024, 1999.

Greely, H. T., "Human Genome Diversity: What About the Other Human Genome Project?" *Nature Reviews Genetics,* 2:222–227, 2001.

Hoyle, R., "US National Bioethics Commission: Politics as Usual?" *Nature Biotechnology,* 14(8):927, 1996.

Hoyle, R., "Clinton's Cloning Ban May Threaten Genetic Research," *Nature Biotechnology,* 15(7):600, 1997.

Hoyle, R., "Arrogance on Human Cloning May Pose a Threat to Biotechnology," *Nature Biotechnology,* 16(1):6, 1998.

Jaenisch, R., and I. Wilmut, "Don't Clone Humans!" *Science,* 291: 2552, 2001.

Juengst, E., and M. Fossel, "The Ethics of Embryonic Stem Cells—Now and Forever, Cells Without End," *Journal of the American Medical Association,* 284(24):3180–3184, 2000.

Kennedy, D., "Two Cheers for New Stem Cell Rules," *Science,* 289: 1469, 2000.

Klugman, C. M., and T. H. Murray, "Cloning, Historical Ethics, and NBAC," in J. M. Humber and R. F. Almeder, eds., *Human Cloning,* Totowa, N.J.: Humana Press, 1998, pp. 1–50.

Koski, G., and S. L. Nightingale, "Research Involving Human Subjects in Developing Countries," *New England Journal of Medicine,* 345(2):136–138, 2001.

Lanza, R. P., J. B. Cibelli, and M. D. West, "Human Therapeutic Cloning," *Nature Medicine*, 5:975–977, 1999.

Leon, A. C., "Placebo Protects Subjects From Nonresponse: A Paradox of Power," *Archives of General Psychiatry*, 57(4):329–330, 2000.

Macklin, R. "Ethics, Politics, and Human Embryo Stem Cell Research," *Women's Health Issues*, 10(3):111–115, 2000.

Marden, E., "The Revolution Ignored: A Critique of Cloning Human Beings: Report and Recommendations of the National Bioethics Advisory Commission," *New York University Environmental Law Journal*, 6(3):674–687, 1997.

McGee, G., and A. Caplan, "The Ethics and Politics of Small Sacrifices in Stem Cell Research," *Kennedy Institute of Ethics Journal*, 9(2):151–158, 1999.

Merz, J. F., D.G.B. Leonard, and E. R. Miller, "IRB Review and Consent in Human Tissue Research," *Science*, 283:1647–1648, 1999.

Meslin, E. M., "Engaging the Public in Policy Development: The National Bioethics Advisory Commission Report on Research Involving Persons with Mental Disorders that May Affect Decisionmaking Capacity," *Accountability in Research*, 7(2–4):227–239, 1999.

Meslin, E. M., "The National Bioethics Advisory Commission (NBAC) Report," *Biological Psychiatry*, 46(8):1011–1012, 1999.

Meslin, E. M., "Of Clones, Stem Cells, and Children: Issues and Challenges in Human Research Ethics," *Journal of Women's Health and Gender-Based Medicine*, 9(8):831–841, 2000.

Michels, R., "Are Research Ethics Bad for Our Mental Health?" *New England Journal of Medicine*, 340:1427–1430, 1999.

Miller, F. G., A. L. Caplan, and J. C. Fletcher, "Dealing with Dolly: Inside the National Bioethics Advisory Commission," *Health Affairs*, 17(3):264–267, 1998.

Miller, F. G., and J. J. Fins, "Protecting Vulnerable Research Subjects Without Unduly Constraining Neuropsychiatric Research," *Archive of General Psychiatry*, 56(8):701–702, 1999.

Miller, F. G., D. L. Rosenstein, and E. G. DeRenzo, "Professional Integrity in Clinical Research," *Journal of the American Medical Association*, 280(16):1449–1454, 1998.

Moreno, J., A. L. Caplan, P. R. Wolpe, and the Members of the Project on Informed Consent, Human Research Ethics Group, "Updating Protections for Human Subjects Involved in Research," *Journal of the American Medical Association*, 280(22):1951–1958, 1998.

Oldham, J. M., S. Haimowitz, and S. J. Delano, "Protection of Persons with Mental Disorders from Research Risk: A Response to the Report of the National Bioethics Advisory Commission," *Archives of General Psychiatry*, 56(8):688–693, 1999.

Oldham, J. M., S. Haimowitz, and S. J. Delano, "Reply," *Archives of General Psychiatry*, 56(8):703–704, 1999.

Parens, E., "Tools From and For Democratic Deliberations," *Hasting Center Report*, 27(5):20–22, 1997.

Press, R., "National Bioethics Advisory Commission Draft Report on Human Biological Materials in Research," *Molecular Diagnosis*, 4(2):164–165, 1999.

Reame, N. E., "Remarks Made Before the National Bioethics Advisory Commission, January 9, 1997," *Nursing Outlook*, 45(3):143–144, 1997.

Reame, N. K., "Making Babies in the 21st Century: New Strategies, Old Dilemmas," *Women's Health Issues*, 10(3):152–159, 2000.

"Report on Cloning by the US Bioethics Advisory Commission: Ethical Considerations," *Human Reproduction Update*, (6):629–641, 1997.

"Research Involving Human Subjects: Efforts to Resolve Current Ethical Controversies," *BioLaw: A Legal and Ethical Report of Medicine, Health Care, and Bioengineering*, Special Sections, 2(7–8):279–348, 1998.

Roberts, L. W., "Evidence-Based Ethics and Informed Consent in Mental Illness Research," *Archive of General Psychiatry*, 57(6):540–542, 2000.

Roberts, L. W., and B. Roberts, "Psychiatric Research Ethics: An Overview of Evolving Guidelines and Current Ethical Dilemmas in the Study of Mental Illness," *Biological Psychiatry*, 46(8):1025–1038, 1999.

Robertson, J. A., "Human Cloning and the Challenge of Regulation," *New England Journal of Medicine*, 339:119–122, 1998.

Robertson, J. A., "Ethics and Policy in Embryonic Stem Cell Research," *Kennedy Institute of Ethics Journal*, 9(2):109–136, 1999.

Robertson, J. A., "Human Embryonic Stem Cell Research: Ethical and Legal Issues," *Nature Reviews Genetics*, 2:74–78, 2001.

Roche, P. A., and M. A. Grodin, "The Ethical Challenge of Stem Cell Research," *Women's Health Issues*, 10(3):136–139, 2000.

Russo, E., "A Look Back at NBAC," *Scientist*, 13(20):4–5, 1999.

Shapiro, H. T., "Ethical and Policy Issues of Human Cloning," *Science*, 277(5323):195–196, 1997.

Shapiro, H. T., "Ethical Dilemmas and Stem Cell Research," *Science*, 285:2065, 1999.

Shapiro, H. T., and E. M. Meslin, "Ethical Issues in the Design and Conduct of Clinical Trials in Developing Countries," *New England Journal of Medicine*, 345(2):139–142, 2001.

Shore, D., and S. E. Hyman, "An NIMH Commentary on the NBAC Report," *Biological Psychiatry*, 46:1013–1016, 1999.

Sobel, M. E., "Ethical Issues in Molecular Pathology: Paradigms in Flux," *Archives of Pathology and Laboratory Medicine*, 123(11):1076–1078, 1999.

Solter, D., "Mammalian Cloning: Advances and Limitations," *Nature Reviews Genetics*, 1:199–207, 2000.

Steinberg, K. K., "Ethical Challenges at the Beginning of the Millennium," *Statistics in Medicine*, 20(9–10):1415–1419, 2001.

Thomson, J., "Funding of Human Embryo Research in the US," *Nature Biotechnology*, 17(4):312, 1999.

Vastag, B., "Helsinki Discord? A Controversial Declaration," *Journal of the American Medical Association*, 284(23):2983–2985, 2000.

Walshak, L. G., "Cloning Human Beings: Report and Recommendations of the National Bioethics Advisory Commission," *Journal of Government Information*, 26(6):773–774, 1999.

Weijer, C., G. Goldsand, and E. J. Emanuel, "Protecting Communities in Research: Current Guidelines and Limits of Extrapolation," *Nature Genetics*, 23:275–280, 1999.

Wolf, S. M., "Ban Cloning? Why NBAC Is Wrong," *Hasting Center Report*, 27(5):12–15, 1997.

Woodward, B., "Challenges to Human Subject Protections in US Medical Research," *Journal of the American Medical Association*, 282(20):1947–1952, 1999.

Wright, S., "Coping with Dolly: Scenes from the National Bioethics Advisory Commission," *Politics and the Life Sciences*, 16(2):299–303, 1997.

Young, F. E., "A Time for Restraint," *Science*, 287:1424, 2000.

Zipursky, R. B., "Ethical Issues in Schizophrenia Research," *Current Psychiatry Reports*, 1(1):13–19, 1999.